Mark

Thank you for your major contribution for a successful 1993 in Fuel Supply.

Paul

Jim

Doug

March 24, 1994

Alberta Miners

A Tribute

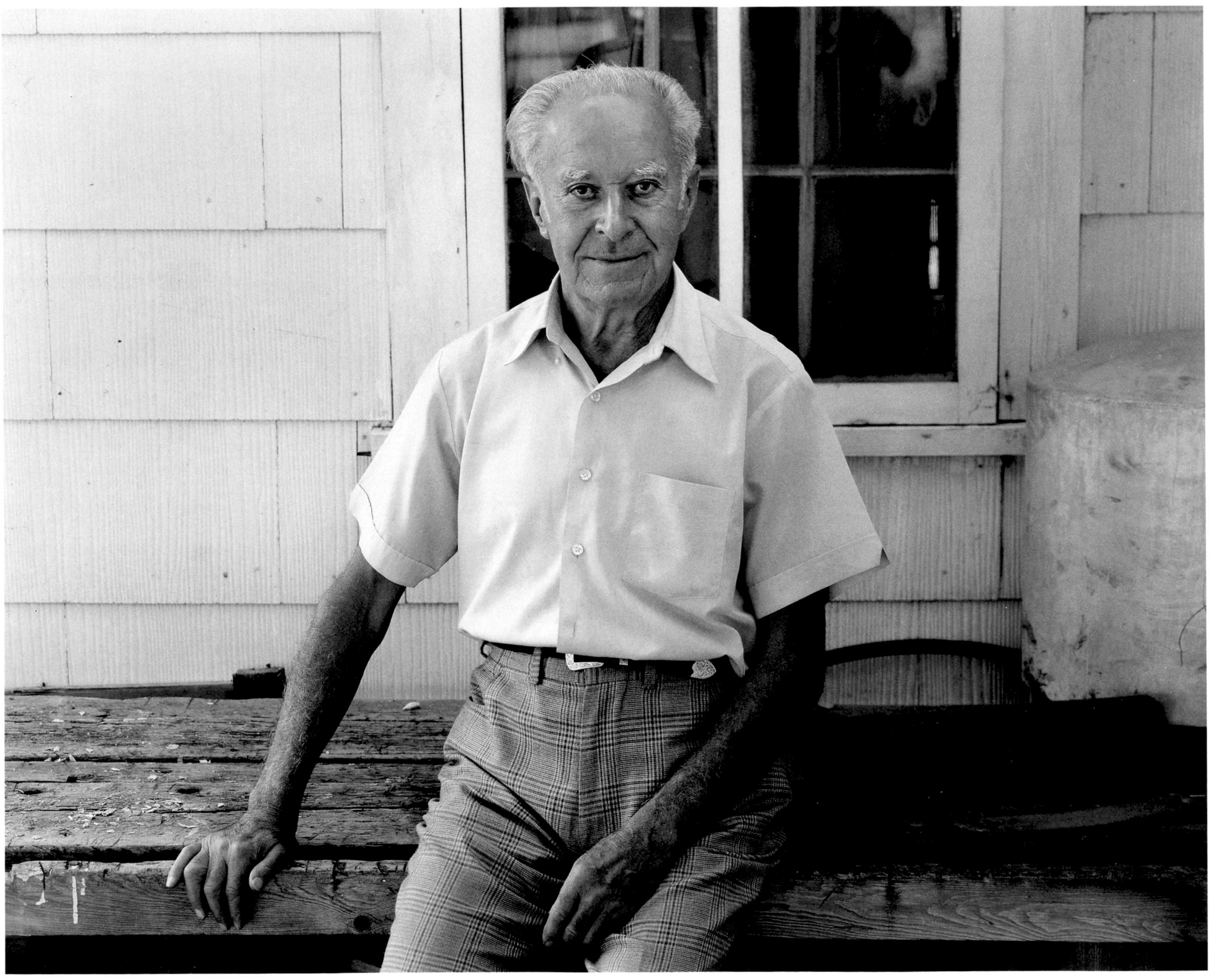

Alberta Miners

A Tribute

Photography and text by Lawrence Chrismas

Foreword by Thomas H. Patching

Produced in collaboration with

The Alberta Chamber of Resources

Cambria Publishing

Canadian Cataloguing in Publication Data

Chrismas, Lawrence
Alberta Miners ~ A Tribute

Includes bibliographical references and index.
ISBN 0-9697023-0-2

1. Miners--Alberta--Portraits. 2. Mineral industries--
Alberta--History--Pictorial works.
I. Alberta Chamber of Resources. II. Title.

HD9506.C23A43 1993 338.2'097123'0222 C93-091325-6

Printed and bound in Canada

Published by

Cambria Publishing
Box 61083
Kensington Postal Outlet
Calgary, Alberta T2N 4S6

Contents

The Alberta Chamber of Resources

As The Alberta Chamber of Resources approaches its 60th anniversary, its members are delighted to have had the opportunity to foster this tribute to Alberta miners.

The photographs in this book are not about the mining companies or the wide variety of equipment existing in the industry — although to be sure the companies, mines, equipment and workers all have an integral role in the success of the mining industry.

Instead, the photographs share the human side of mining. These images portray an intimate cross-section of retired and working miners who continue to influence our daily lives in Alberta.

In fact, upon glancing through this photographic tribute to our miners, it becomes increasingly clear that the mining industry has played a fundamental role in the development of the province, and will continue to do so for its future prosperity.

The Alberta Chamber of Resources is grateful to the members and friends listed on the adjacent page who have contributed to Alberta Miners ~ A Tribute.

Erdal YILDIRIM, Dr. Eng. Sc.

President
The Alberta Chamber of Resources

Sponsors

Alberta Department of Energy, Mineral Resources Division
Alberta Oil Sands Technology & Research Authority
Bruce Hunter
Burnco Rock Products Ltd.
Canada Cement Lafarge (Exshaw)
Canadian Institute of Mining and Metallurgy, Coal Division
Cardinal River Coals Ltd.
Coneco Equipment Ltd.
Consolidated Concrete Limited
Echo Bay Mines Ltd.
Finning Ltd.
Fording Coal Limited
Garritty & Baker Drilling (1979) Ltd.
Geophotographics
Imperial Oil Resources Limited
Luscar Ltd.
Monenco Agra Inc.
Nordegg Lime Ltd.
Northward Developments Ltd.
PanCanadian Petroleum Limited
Sam and Gerry Shachnowich
Sissons Mines Ltd.
Stanley Industrial Consultants Ltd.
Suncor Inc. Oil Sands Group
Syncrude Canada Ltd.
The Coal Association of Canada (History & Heritage Committee)
TransAlta Utilities Corporation
TransCanada PipeLines Limited
Transwest Dynequip Ltd.
T.A. Klemke & Son Construction Ltd.
Wajax Industries Limited
Wirtanen Electric Ltd.

Mine Sponsors

Century Coals Limited (Atlas Mine)
Consolidated Concrete Limited (Cadomin Quarries)
Fording Coal Limited (Genesee Mine Operations)
Fording Coal Limited (Whitewood Mine)
Forestburg Collieries (Paintearth)
Luscar Sterco (1977) Ltd. (Coal Valley Mine)
Obed Mountain Coal Ltd.
Syncrude Canada (Oil Sands Mine)

Foreword

Little is known about the first people who made use of Alberta's minerals. Most of what we do know of them comes from study of their stone knives, scrapers, projectile points, and the quarries and sites where these tools were made. Production of stone tools and weapons continued until the Europeans arrived, only a few hundred years ago. Except for the recovery of tarry material in the north-east part of the province for patching canoes, this was for thousands of years the sole mining activity in Alberta. Further, there is no evidence that the native people made use of coal that was exposed along the banks of several rivers and mountain slopes.

In 1792, Peter Fidler was the first man to report the presence of coal in Alberta, on the banks of Rosebud Creek. He related that he brought coal into his tepee for fuel, and that this action apparently horrified his Indian companions, since western tribes had a taboo against its use.

Nothing more was reported about coal in Alberta until Dr. James Hector with the Palliser expedition in 1859 noted the burning of coal beds along the Red Deer River. Dr. J.B. Tyrrell's studies in the following couple of years revealed the existence of very large quantities of coal in several parts of the province. It is possible that some coal might have been taken from seam exposures during this period by fur traders and others who were starting to move into the province in the middle of the 19th century. But such mining has not been reported.

In 1857, Dr. Hector saw a small amount of gold from the North Saskatchewan River. Two years later, men on their way to the Caribou gold fields passed through Alberta and, in 1860, prospectors reported gold discoveries in the Rocky Mountain House area. A promotion in eastern Canada of opportunities in the Saskatchewan River Gravels promptly drew several parties of men to Edmonton. Some went on as overlanders to the Caribou, while others spread out about the province in a mini gold rush, generally with disappointing results. Rumours and promoters are not new in the mining game.

Sporadic washing of river gravels for gold has continued in the province since those days. There have been at least five attempts at gold dredging in the Edmonton district. During the hungry thirties, placer miners eked out a living panning along the rivers, sometimes living in dug-outs and tents on the banks. Even today, small amounts of gold are recovered as a by-product of some gravel-washing operations.

Nicholas Sheran opened the first mine in Alberta. In 1870, he came north from Fort Benton, Montana, hoping to find gold. On seeing coal outcrops near Fort Whoop Up, he decided to start a mine and sell coal to local ranchers, hauling it by ox team to Fort Benton. He began his first underground entry in 1872, near the site of Lethbridge's high level bridge.

Sheran's venture was not successful and he died in poverty ten years later. But by then, eastern businessmen had become interested in the district's mining potential and Sir A. Galt formed a company to establish mining operations there. Originally, production was expected to supply the ranchers and incoming settlers, but as the CPR stretched west across the prairies, it needed fuel for its locomotives. A narrow-gauge railway was laid to take coal from Lethbridge to Dunmore, near Medicine Hat, connecting with the CPR mainline in 1885.

Opening of the prairies to settlement was restricted by the need for winter fuels. Many homesteaders travelled distances to coal outcrops where they dug out coal, sometimes unofficially, for the coming winter. For example it is related that C.O. Card and two others set out in 1887 to find coal for their new settlement at Cardston and in June, "found a small vein of coal — which provided all with greater comfort." As the prairies were being settled, many mines were opened, especially around Edmonton, Drumheller and Lethbridge, to provide coal for local and eastern markets.

Meanwhile, the railways moved westward. The great coal deposits in the foothills and mountains were discovered and mines were opened by large mining operations, many financed by eastern or foreign capital. These mines produced fuel for domestic or railway markets, depending on the coal's properties.

The rapid growth of coal mining in Alberta is shown by production figures. The first recorded tonnage in 1886 was 43,000 tons; this tripled four years later. By 1900, production reached 311,000 tons, and the figure grew by ten times in the next ten years. By mid-1920, coal output doubled again and there were several hundred mines operating in the province.

As the number of coal mines in the province increased, so did the need for skilled and experienced miners, as well as for other men who could be trained and added to the work force. Many early mine supervisors came from mines in the United Kingdom, while large numbers of other employees were drawn from eastern and western Europe. Some came equipped with good mining experience, while others had only a willingness to learn and work.

Many early mining communities in the foothills and mountain districts were isolated, their only means of access by railway or wagons. Because they were on their own, the people in these communities also provided their own services and entertainments. Almost every mine or mining town could boast of a hockey and baseball team, while libraries and brass bands — and pubs —

were strongly supported. A variety of languages could be heard in the mine drys, on the streets and in the school yards as the new families settled in.

It is to the credit of many mining families that their children got a good education. Young people who grew up in the mining towns have often distinguished themselves in music, science and engineering, medicine and law. (During my tenure at the University of Alberta, it was my good fortune to know quite a few young men from mining communities who had come there to get degrees in mining engineering.)

The character of coal mining in Alberta has changed greatly since the 1939-45 war. Before the war, almost all coal came from underground mines, which employed large numbers of manual workers. Now, with the exception of one underground mine, most Alberta coal is produced by large strip or open pit mines, using mechanical excavators and haulers, and employing far fewer employees to operate and maintain the equipment. The introduction of diesel locomotives on the railways soon after the war forced the closure of most mountain and foothill mines, ending the employment of hundreds of miners. Concurrently, the extension of natural gas pipelines for domestic heating caused the demise of most underground prairie mines.

On the other hand, the extension of electrical power through the province necessitated the opening of several large strip mines to supply the coal to electrical generating stations. In the 1970s, an expansion of export markets for coaking coal also supported the opening of several large strip and open pit mines in the mountain areas. Consequently, the total tonnage of coal currently mined is now larger than before the war, although far fewer mine workers are employed and the nature of their work has changed. The small mining communities that were located usually within walking distance of the mine shaft or tipple have disappeared. Most miners now live in larger towns, served by good roads, at some distance from their places of work. There they enjoy the same living conditions and pastimes as the other residents of the communities.

While coal mining was Alberta's dominant mining industry in past years, it is now matched in size by the two huge oil sands mines and extraction plants near Fort McMurray. The first of these open-pit mines, the Suncor Mine, was opened in the mid 1960s, employing large bucketwheel excavators that had never before been used in the country. The second, the Syncrude Mine, was brought into production in the late 1970s and was even larger. It uses very large draglines, bucketwheels, loaders and conveyors. Much of the equipment in these mines is unique in size and nature, and most of the work involves the operation and maintenance of this equipment. Although the employees are mine workers, their jobs do not much resemble what used to be considered traditional mine labour. Employees for both mines and extraction plants reside in the fully modern city of Fort McMurray. Although it is separated many miles from other major communities, it does not resemble the small one-industry mining towns of former years.

In Alberta, coal and oil sands generally come to mind when mining is mentioned. However, several other mining activities also play important roles in the province's history and economy. These include stone quarrying, the production of limestone, clays and shales, and the operation of numerous sand and gravel pits. We see or use these products almost every day.

Only a limited amount of stone quarrying for building purposes is now carried out in the province. Several small quarries within the mountain valleys have produced 'Rundle Rock,' the stone that gives character to many of the buildings in Banff Park. No longer visible or active, there were a number of sandstone quarries within the Calgary city limits from about 1886 to 1915. Stone from these quarries went into so many of the fine buildings in Calgary that it was at one time called "The Sandstone City." These quarries also produced stone used in various buildings throughout Alberta.

Many of the quarrymen and stone masons who worked on these buildings were immigrants who had learned their trade in Scotland. In the years of peak activity, several small communities of stone workers grew up near the city quarries.

The massive limestone formations in the mountains provide an ample source of stone for the province's cement and lime products. The quarries in the Crowsnest Pass, Exshaw and Cadomin areas have operated continuously for many years. Quarries at Exshaw, where the mountains meet the foothills, sustain the largest cement-making plant in western Canada, as well as a smaller lime-burning plant. The limestone quarry at Cadomin ships the crushed rock by rail to Edmonton, for processing. In Cadomin itself, limestone quarrymen now make use of houses left by the town's former coal miners.

In the early years of Alberta's settlement, most of the major buildings in the province were made of brick, as were the private homes of 'people of substance,' except where stone was available. As the population grew, there was a corresponding need for brick yards and plants, and between 1906 and 1913, 32 plants were established at various places around the province. Many more brick yards or companies show in records, but some never got into production. Plants were located near a supply of good clay or shale, water and fuel, and preferably close to markets. Some of the shales in early years were mined underground, but later all shale and clay was taken from open pits, using conventional excavating equipment. Redcliff and Medicine Hat became the principal sources of brick in the province because of the cheap natural gas and excellent clays in these districts.

Regrets are sometimes expressed about the disappearance of the 'good old days' when life was simpler, and people lived and worked closer together, and knew each other better. Certainly there was a great deal of fellowship between the men who worked together in the mines. Sometimes also there was rivalry and strife, but generally there was a common bond between those who shared work. People living in a mining community were usually considered neighbours. Former residents of places like the Coal Branch or The Pass recall the dances and picnics, and speak fondly of the good life when their families were growing up. Although, at the same time, many would no doubt be reluctant to go back and start it all over again.

Many of those who worked in the mines in former years are now gone and their memories and mining lore are being lost. In this respect, the photographs and stories obtained by Lawrence Chrismas help to preserve some memories of a mining life, and to honour those who have worked in Alberta mines.

The focus of this documentation is on the average worker, not on company officials or executives. However, a mining operation requires a full range of workers, from novice to chief executive, to handle its operations. Accordingly, an attempt to show the industry's variety of positions has been made in this book.

The photographer's experience goes back to 1979 when he first photographed Atillio Caffaro, a miner with 50 years at Canmore Mines. As the chief mechanic at Canmore, Atillio was one of the hard-working and successful people who made Alberta their adopted home. That photograph, the frontispiece of this book, was the beginning of an ongoing passion for photographing and recording the histories of retired miners in Alberta and elsewhere in Canada. Later, the project expanded to include working miners — a perspective that graphically demonstrates the changes that have transformed the industry over the years.

Thomas H. Patching

Edmonton, Alberta

Photographing miners in Alberta and across Canada has been a major obsession of mine during the last 14 years. The charisma of these people, their diversity of character and their personification of mining history keep me returning to my subject, either to search for new contacts or to revisit old friends.

The Alberta Miners project began in 1979, when I first met Atillio Caffaro and friends — the day after Canmore Mines announced its closure after 93 years of operation. The announcement prompted Atillio to pour out his thoughts on his own 50 years of mining in Canmore.

I listened, expecting to hear stories of the downtrodden mine worker whose life was filled with poverty and injury. But this was not the case. Atillio was genuinely distressed by the closure of the Canmore Mine, as were his former co-workers. And as I continued to photograph and interview these people, I found that most had enjoyed their lives in the mines, although they were also envious of their modern-day counterparts with their high earnings and their new pickup trucks.

In fact, the opportunity to visit these old-timers in the relaxed atmosphere of their homes, and to hear their recollections of the mining industry, was one of the most rewarding aspects of the project. Their stories proved to be personally enlightening, shedding some light on the experiences of my own grandparents who came to Alberta at the turn of the century.

Documenting contemporary working miners was another matter — but one that demanded the same level of respect. For the most part, I photographed these highly skilled operators of expensive and sophisticated machinery in their working environments. Many workers took short breaks from their tight schedules to pose and talk to me, their professionalism in sharp contrast to the laid-back reminiscences of the retired miners.

But one aspect of the industry continues to abide through both generations of miners — the companionship of the workers. Although today, mine employees are often scattered across their working environment, the old camaraderie still exists when workers get together at shift change or coffee breaks.

During my first five years of photographing miners, I concentrated on portraits of retired miners; later, the project evolved to include working miners at operating coal mines.

When the opportunity arose to produce a book for The Alberta Chamber of Resources, I welcomed the chance to photograph other types of miners, those individuals involved in limestone quarries, the oil sands, and other mining ventures. As well, the comparisons between retired and working miners became more vivid — the most obvious difference being that many old-time coal miners worked underground, whereas today only two underground facilities exist in Alberta: the Smoky River Coal Mine and the Underground Oil Sands Test Facility.

The people in these photographs are not candidates for the Order of Canada, but within their families — and in my eyes — they truly are heroes. I have tried to photograph them accordingly.

For the most part, I have avoided the rigid formality of portraiture and tried to impart a sense of the sitter's environment and his position within it. When asked to explain my style, I say that I often think of myself as a snap-shooter with a large format camera.

Provincial laws in Alberta require that persons working in mining operation abide by the safety regulations. In most situations, workers must wear hard hats, protective eye wear and steel-capped boots. If this protective gear is absent from selected photographs taken at operating mines, the photographer accepts the responsibility for requesting the removal of hats and glasses.

Lawrence Chrismas

Calgary, Alberta

Dedication

To my heroes,
wherever you have gone,
the old-time miners

and to

the preservation of our mining
history and heritage.

Southern Alberta

Locations of photographs and mines

Mines/Companies

Summit Limestone Quarry,
Summit Lime Works Ltd., Hazell

Korite Ammolite Quarry,
Korite Minerals Ltd., Welling

I-XL Clay Quarry,
I-XL Industries Ltd., Medicine Hat

Continental Limestone Quarry,
Continental Lime Ltd., Kananaskis

LaFarge Limestone Quarry,
LaFarge Canada Inc., Exshaw

Rundle Rock Quarry,
Rundle Rock Building Stone (1980) Ltd.,
Harvie Heights

Thunderstone Quarry,
Thunderstone Quarries Ltd., Dead Man's Flats

Springbank Aggregate Pit,
BURNCO Rock Products Ltd., Calgary

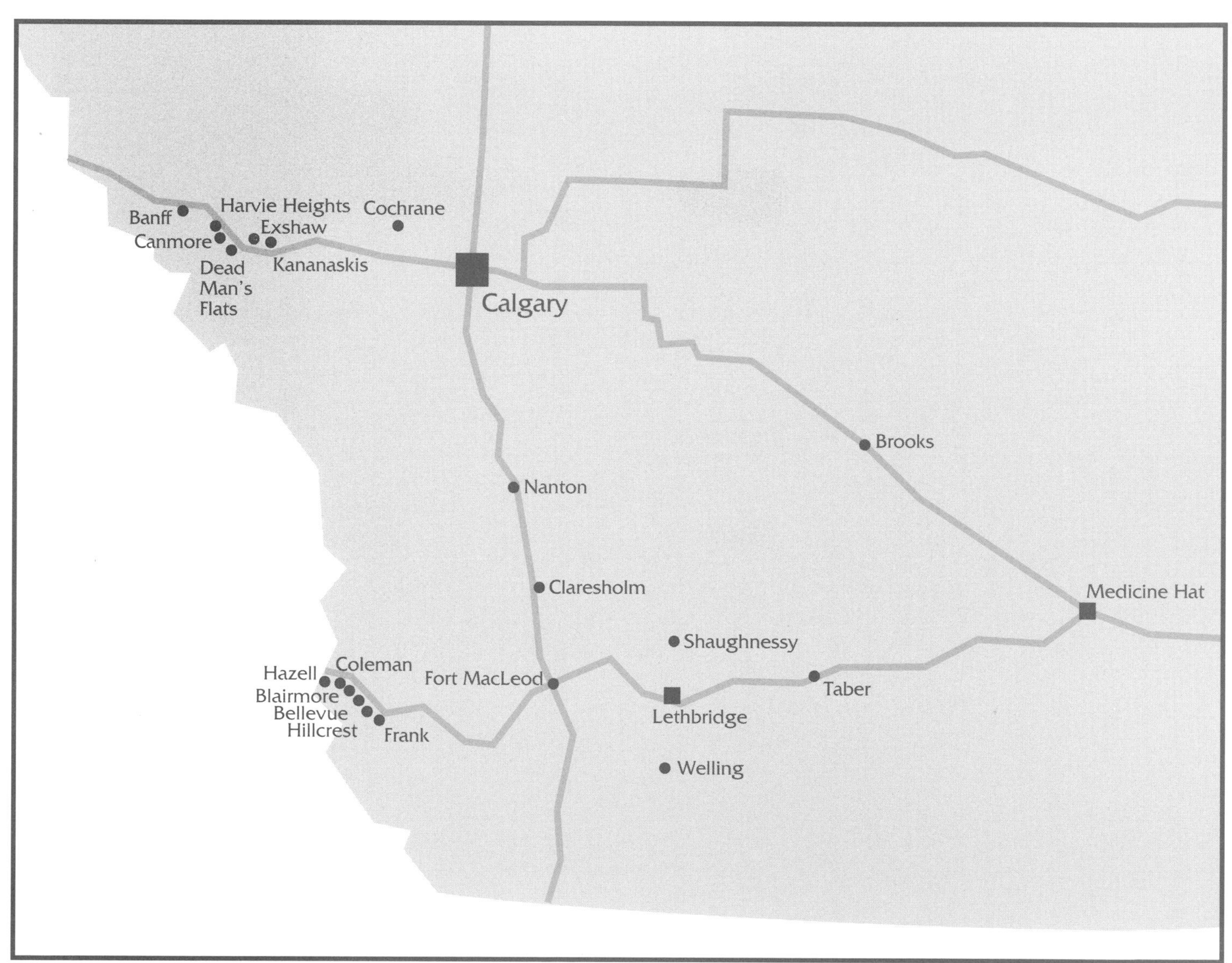

Summit Limestone Quarry, Hazell

Yoho In 1906, my dad moved to Cumberland, British Columbia from Japan. He was a coal miner there until 1938, when they wouldn't hire any more Japanese in the coal mines. They evacuated him from the coast to the Slocan Valley — he was 70 years old. I was 18. They wanted to send me to a road camp, and when I refused, they chucked me into an internment camp for four years. In the end, I came to the Crowsnest Pass because my family was here. About a dozen Japanese families lived and worked here in the '40s.

We have seven children now and there are still four around here. I've got one son who still works at Westar Coal, and another one who's working here.

My first job at Summit Lime was pushing wheelbarrows of limestone rock, used to make calcium. We'd go up into the quarry and use a 16-pound hammer to break up the limestone so you'd have the right size pieces for the kilns. That's how backward it was in 1947.

Later, when we got a big order of sugar factory rock, the company got some crushing equipment.

The quarry is much bigger now than it was in the old days. We used to produce about 40 tonnes of lime per day, whereas now we can produce more than 200 tonnes a day.

I've been a kiln operator or fireman for over 30 years now. If I didn't enjoy it, I wouldn't be here.

Yoho Kimoto, Kiln Operator

Summit Limestone Quarry, Hazell

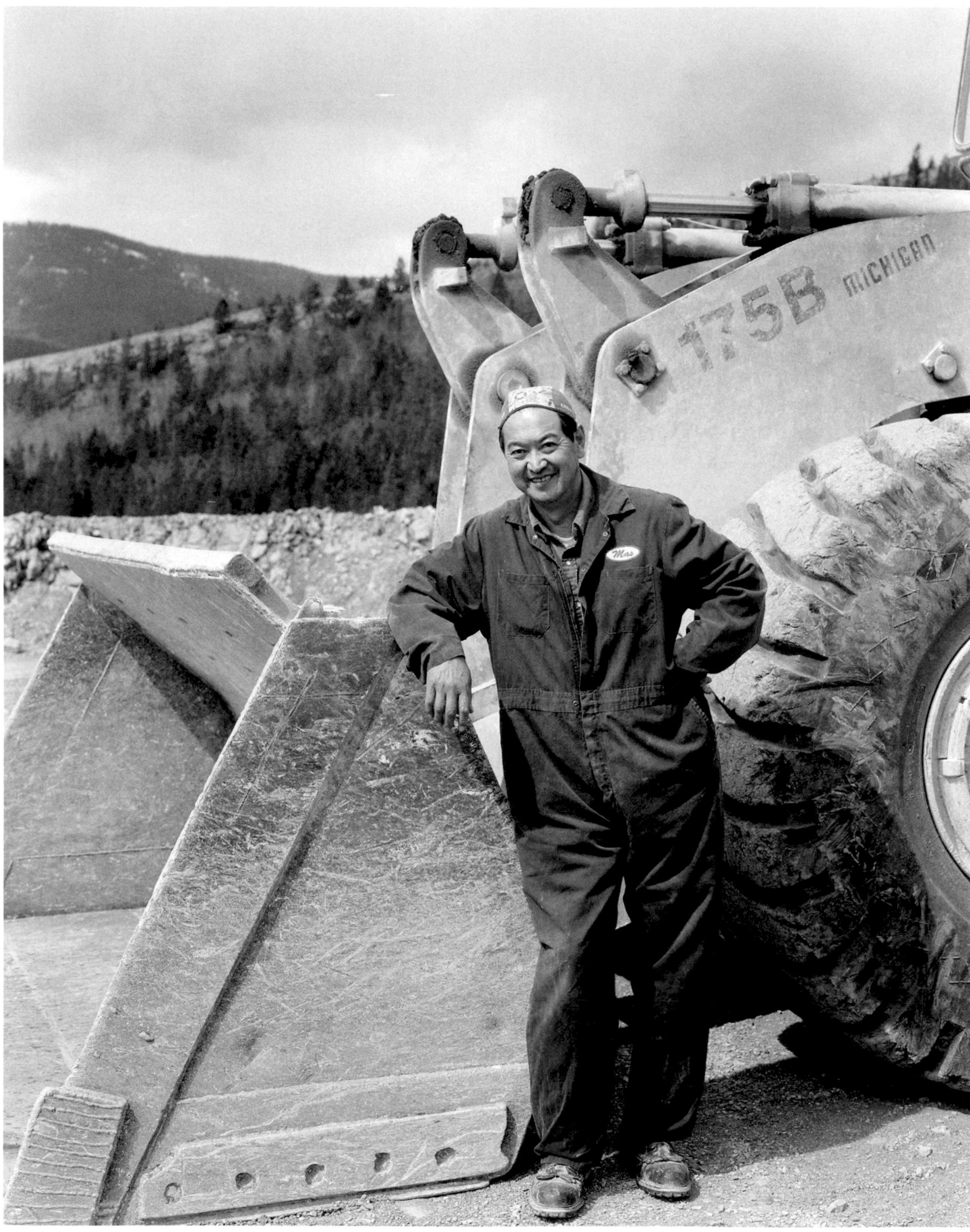

Mas Kimoto, Loader Operator

In the late 1880s, two Italian plasterers from Toronto first quarried limestone by Crowsnest Lake near the Alberta/British Columbia border. The two men built a small masonry beehive kiln with the aim to export lime to Chicago, where a comparable plastering product was not available.

In 1903, a Welsh plasterer named Hazell, recognized the limestone's special properties as an easily spreadable lime putty — and he purchased the property. The Hazell family owned Summit Lime Works until 1991, when it was purchased by Continental Lime. The year 1993 marks the 90th anniversary of Summit Lime.

Earlier, another Crowsnest Pass company established a quarry and lime kilns between Frank and Blairmore, at the edge of the Frank Slide. Its short lifespan, from 1912 to 1922, was blamed on poor markets.

Limestone is quarried by the retreating bench method, in which 30-foot-high benches are dug down from the top of the hill. The product is then drilled, blasted, loaded and trucked to the plant. Today, Summit production averages about 25,000 tonnes of lime annually.

The general term lime is given to burned or calcined limestone and its secondary products of slaked, hydrated and quick lime. Within the kilns, high temperatures completely break down the limestone to produce lime and release carbon dioxide.

In Summit's early days, workers hand-loaded materials into kilns. Local coals were gasified and the volatiles released from coal were piped into the kilns. In current operations, the limestone is processed in continuous natural gas-fired vertical kilns.

When the quarry first opened, transportation was so limited that the company built employee homes next to the plant and quarry. Workers had literally only to fall out of bed to go to work. Although six houses at the quarry site are still occupied, most employees now live in Coleman.

Rick Beadle, Driller and Norm Cervo, Blaster

Erwin I was born in Poland in 1917. My family came to Canada in 1922. My father worked at the Bankhead Mine in the briquette plant, the last year the mine operated. Then he went underground at Michel, British Columbia. I started with my father as a contract miner at Michel. In all, I worked 31 years underground and 14 on the surface.

Coal mining was more or less for family. If your father was a miner, you'd go down to the office every morning before they started and look for a job. Mining coal as a contract miner was good work. There was always something about it that was a challenge.

You can always tell a good miner when you see one — just look at his tools. If he has ugly-looking tools — dull and everything — that's the kind of miner he is. If he has sharp tools, *he's* sharp. I still have 15 or 16 picks in my basement. To use a pick on coal took knack, not muscle. You could feel the gas work the coal once you got started.

I got laid off after 45 years. How could they lay me off after 45 years?

Easy — because a person is no good any more.

Erwin Spievak, Retired Underground Coal Miner

John I was born in Scotland and came to Canada in 1926, when I was 12 years old. I'm from a family of coal miners. My grandfather was a miner in Scotland and my mother worked in a large shaking-and-screening plant. She was on the tables, picking the slate out of the coals.

During the '30s, I got my architect's diploma from Tech. I worked for seven years at the International Mine until they found out I had a diploma — then they shoved me into the office. I spent the first ten years of office work changing the International plant from a dry to a wet washery. I built the briquette plant in the '50s when the railway switched to diesel. Then I ran the plant for a year. I should add that I also have a fireboss ticket and a provisional mine manager's certificate.

I married a coal miner's daughter from Inverness, Nova Scotia. Her dad started in coal mining when he was nine, cranking a fan by hand for ten hours a day. He'd fall asleep on the job until a miner would come down and rap him on the head with a pick handle. He ended as one of the best contract miners in Coleman. There's a technique to mining coal, and he knew it.

In 1978, our son (who has his fireboss ticket) was very active working for the mine rescue team. He took the Alberta championship for Coleman Collieries. Then he won the National in Nova Scotia.

I have one son working at the Line Creek Coal Mine in British Columbia, not far from here.

Marie and John Kinnear, Retired Coal Mine Surveyor

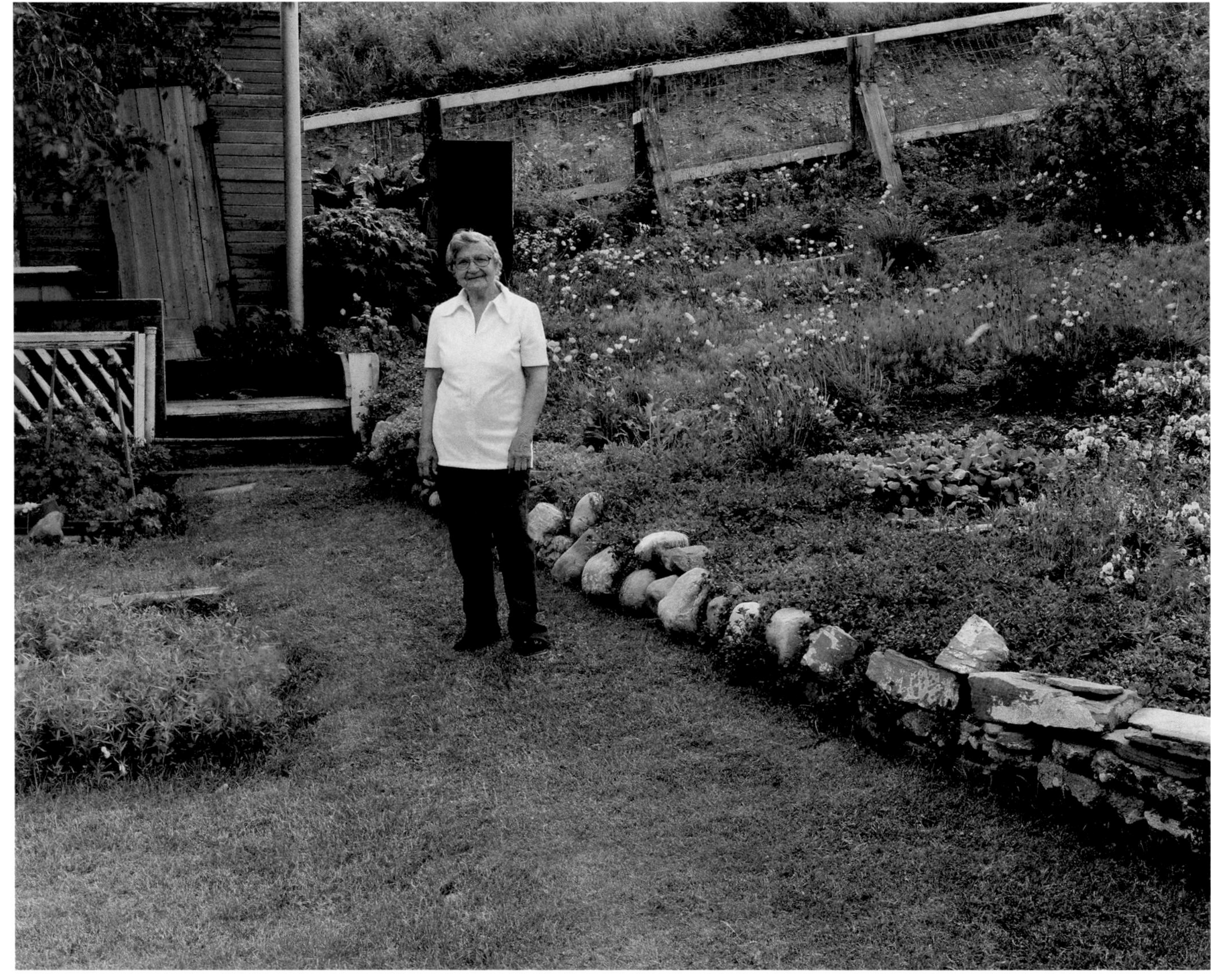

Kay Yates

Kay I never did regret coming to Canada. This is a great country, the best in the world....

I was born in Lancashire, England in 1899. I married Jim Yates, a coal miner from Michel, British Columbia in 1918, in England, when he was overseas during the war.

In 1919, when the war was over, we returned to Michel, where Jim resumed working as a coal miner. A year later, we went to Coleman and the McGillivary and International mines, where Jim got better wages. We built this house in Coleman in 1922. When the mines started to slow, you could buy a house for $2,000.

There was an explosion at the McGillivary Mine when he was there and for a short time, I thought he had been killed. But he was able to get out. Several miners lost their lives in that accident.

I was a witness at the trial of a local rum runner, Picariello, and Mrs. Lassardro. They were both found guilty and hanged for shooting a local police officer. Picariello was a good man. He was set up.

My husband Jim used to say when they need the miners here again, they'll all have moved away. He died in 1968.

Willy I've been working for over 40 years. It's a long time to pack a lunch bucket. I can't even look lunch meat in the face anymore.

My dad came to Blairmore from Poland, and then went to Detroit. After I was born, in 1926, we returned here, where my dad worked in the West Canadian Collieries Mine. I spent all my life in Blairmore.

I was 18 when the money attracted me to work in the West Canadian mine. I started out on haulage, bucking coal, and from there I went to pipe-fitting for about nine years. Yeah, they wouldn't let me off that damn job.

My job was to put the pipes in the work areas so the miners could run their air picks. When the pillars were finished, then I would go in and extract the old pipes before the roof caved and they were buried. It wasn't a bad job unless you were the last guy in the pillars and the roof was cracking. Oh yeah! I ran a few times!

Later, I worked on the Grassy Mountain Strip operation until it closed; then I went on to Coleman Collieries, where I worked nearly 20 years. Now, I'm working for Byron Creek at Corbin.

Willy (The Cobbler) Habdus, Former Underground Coal Miner

Dale I'm a third generation coal miner, born in Coleman in 1949. My father, who was born in Corbin, worked in three different mining towns in the Crowsnest Pass area.

While I was at university, I spent my summers underground at the Vicary Mine in Coleman. And when I'd completed my degree in Anthropology at the University of Lethbridge, I took a job in the underground hydraulic mine at Sparwood.

My dad thinks underground mining the old-fashioned way was the best. However, I think the hydraulic mine is the best technology.

I don't particularly like coal mining, but I guess I will probably do it for the rest of my life. It's okay.

John My mother had six brothers who were killed in explosions in Welsh mines. She said she'd skin me alive if I ever went into the mine.

So when I went into the mine at age 17, I told her I was working on the tipple. I was in the mine nearly a year before she found out about it.

I was born in Wales in 1923, and my family came to Canada in 1930. I started in the Hillcrest Mohawk Mine and worked there until it closed. After Hillcrest, I worked in Coleman and Michel. When everything began to fold up in the coal industry, I went to work in the hard rock mines. I thought coal mining was finished.

But when I returned, they were mining coal at Vicary Creek in Coleman. It was even better than before because the company guaranteed the men a five-day week and introduced continuous mining machinery.

In coal mining, every day is different; that's what I like about it. You can't say that it's boring.

Let's put it this way: if you could rerun the whole damn thing like a tape, I'd run it all back the same way, the same bloody way!

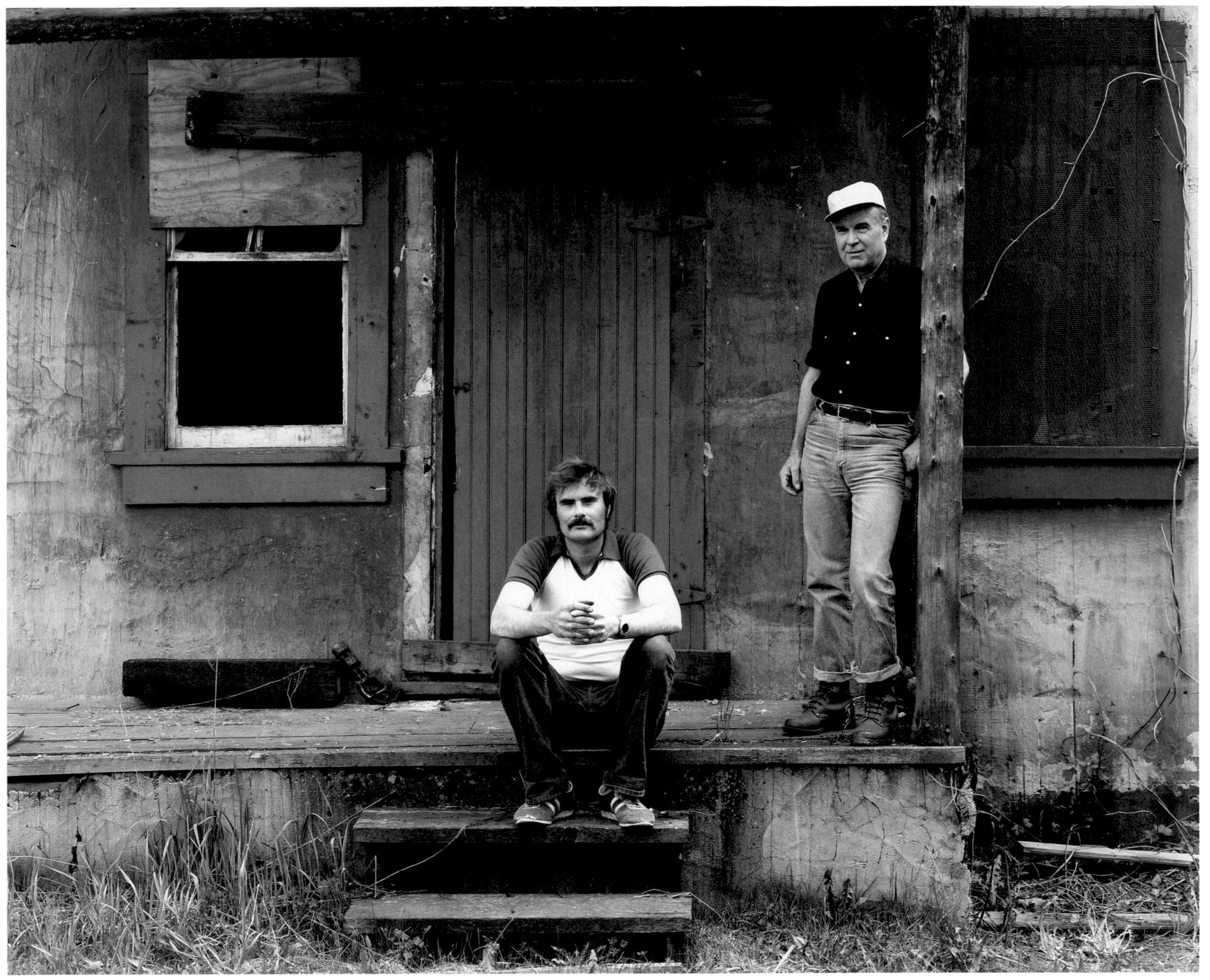

Dale Montalbetti and John Fry, Underground Coal Miners

Albert In 1908, the year I was born, my family came to Glace Bay, Nova Scotia from England, and from there went to Bellevue. I was 16 years old when I started as a driver's helper at $3.50 per day — that was my job for two years. After that, I dug coal for 25 years. Then I was put up on the hill. I spent 37 years in Bellevue and six more at the Vicary Mine in Coleman.

My dad always said the Bellevue Mine was the best in North America — and he should have known, he worked in all kinds of mines.

I was put up on the hill here during the war. That was the most foolish move I ever made in my life. I had gone to night school and received my certificate for mining competency. Now, because I had charge of the whole mine, I had to work Saturdays and Sundays, and the pitbosses would call me in the middle of the night when there was a cave-in and they "couldn't start the pumps." So I used to have to get up in the middle of the damn night and show them how to start the pumps!

When I went to Vicary, I had to pay $50 to join the union — I was lampman at the time. Six months later, they asked me to go in the mine as a fireboss, so I had to get out of the union that I had just paid to get in!

As a hobby in the summer, I've been working as a guide at the Leitch Collieries Provincial Historical Site.

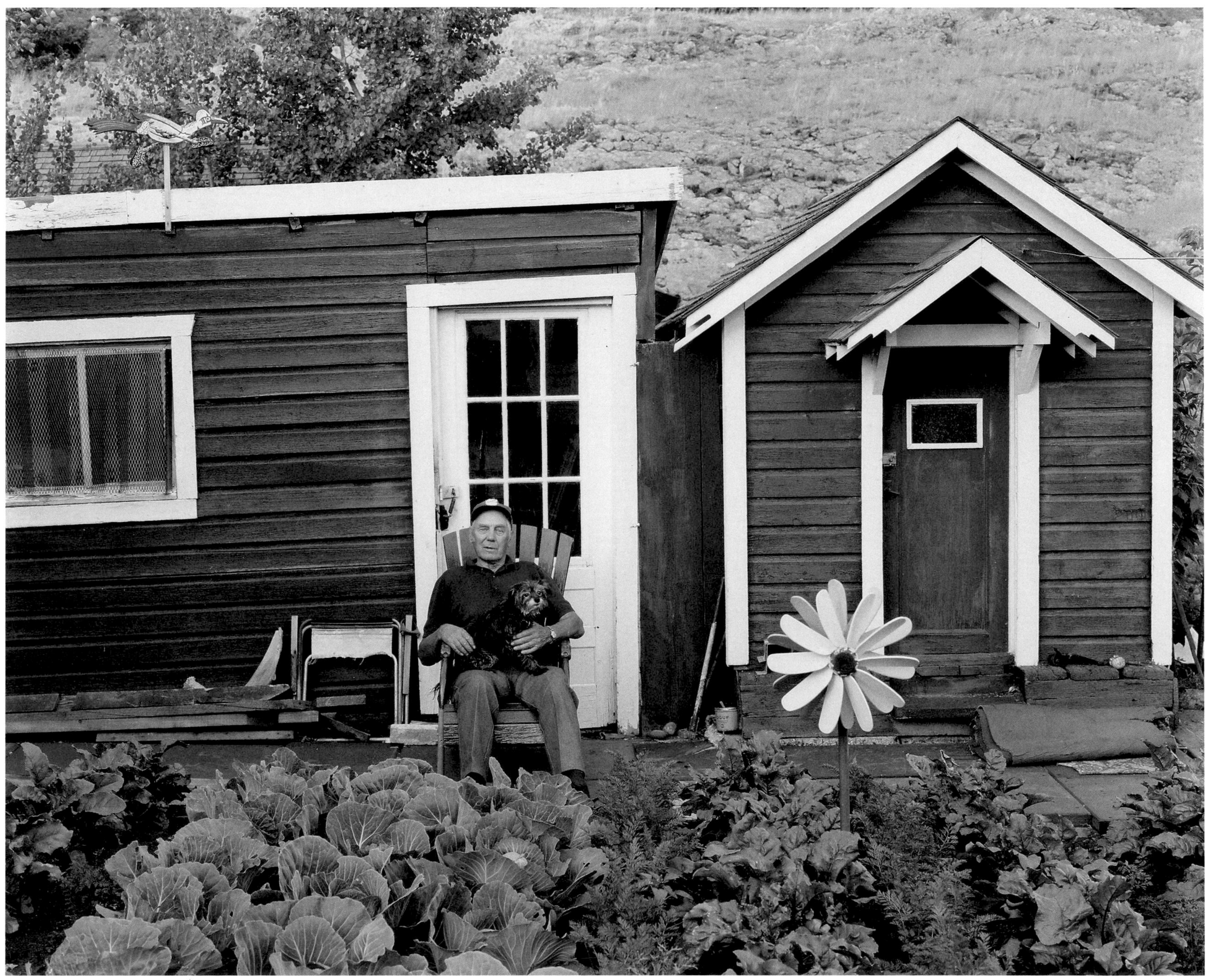

Albert Goodwin, Retired Underground Coal Miner

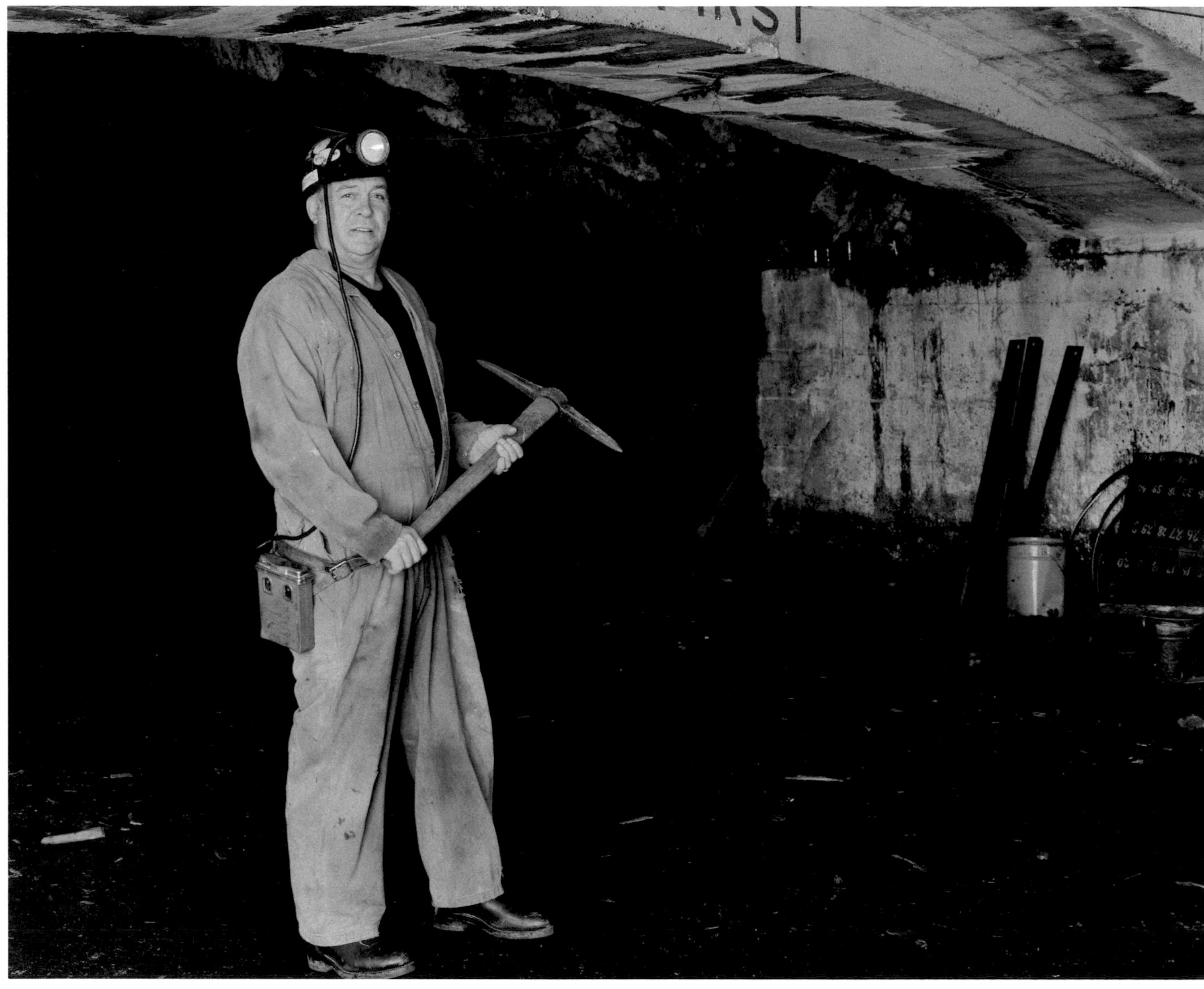

Roy Lazzarotto, Underground Coal Miner

Roy When I go into the mine, I am born again. I always see something different. Different than other people see. I see the formation of rock and coal. I would never trade my mining life for anything else.

I was born in Italy in 1931. I travelled to Lethbridge in 1949 to visit my uncle, learn a little English and see the country, before I went back to Italy. But it didn't work out that way. I met a little girl in the Pass and we got married. She was from Drumheller, and her father was a blacksmith.

I started in the Bellevue Mine as track layer, and ended up doing a variety of jobs until it closed in 1957. Then I went to Coal Creek, and the Michel and Natal mines in British Columbia. I worked on the first continuous mining machines in 1960, and I started the Hydraulic Mine. I returned to working in Alberta when I began with Coleman Collieries. In all, I worked 35 years in the mines in Canada.

For the past three years, I have been looking after the underground restoration work at the Bellevue Mine. We've opened it for public underground tours, and have a small museum there.

I really liked the Bellevue. It was a gentleman's mine.

Victor There are many good things about mining if you're not in a hazardous situation. You never know when it's going to cave. It cracks and breaks and it makes you run. Then you've got to go and dig again, and then run out again. Well, some guys were crazy being brave, and some guys were cowards. You know, a live coward is better than a brave guy who gets killed.

My father came here in 1913, from Russia. He was a coal miner for 45 years, and worked mainly in the Mohawk Mine as a timber packer boss. I was born in 1925 in Hillcrest. I started bucking coal in the Mohawk Mine with my buddy, John Fry when I was 16 years old. This work was followed by jobs in the Hillcrest Mine and the McGillvery Mine in Coleman.

I bought this house in 1942 for $800.

In the early '50s, we were off work more than we were on work. In 1959, I went to the King Gething Mine in Hudson Hope, British Columbia. Then, in 1969, I went to Grande Cache as a fire boss, and became the underground foreman. I dug 78 adits for coal exploration. I was glad to be at Grande Cache. It's lovely country. I looked forward to going to work every day. I was making good money and working six or seven days a week, sometimes.

You know, a coal miner is just like a fisherman with his fish tales. In the bar, we dig more coal than in the mine, because everybody brags about what they do. I know one thing: there is no coal miner who knows everything. One never knows too much, there's always room to learn.

Victor Belik, Underground Coal Miner

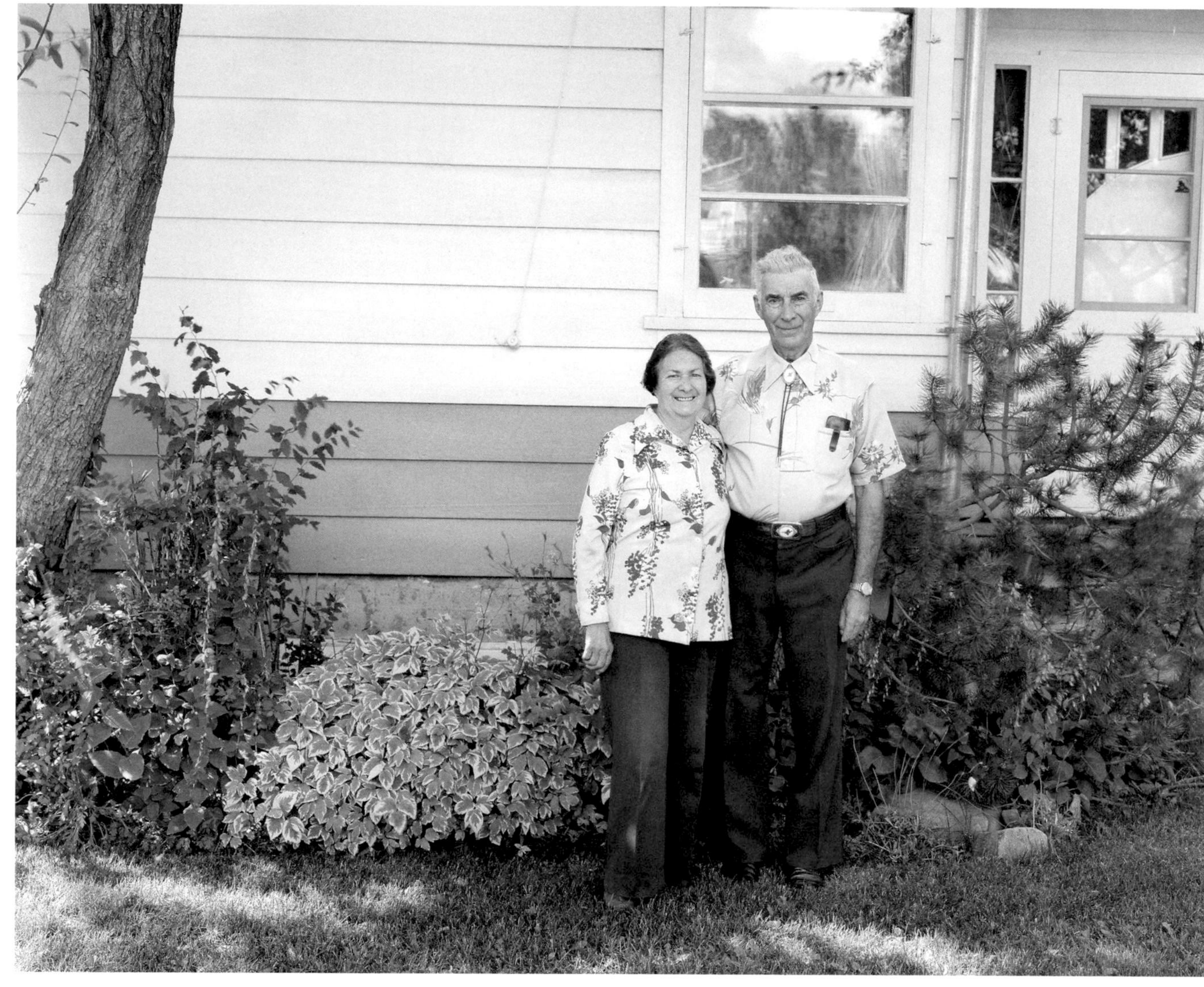

Ethel and John Curry, Retired Coal Mine Safety Coordinator

John It was automatic that after you had gone through school, there was no thought of going any further. Grade eight was it. The next thing was a job.

I started in coal mining more through necessity than choice. It was the only thing that was feasible at the time.

But still, while shuffling around in the coal industry, I worked my way pretty well from the bottom to the top.

When I was 15, I got my first job at West Canadian Collieries in Bellevue. In 1924, I started work in Hillcrest and stayed there until 1939. I couldn't get a job at Bellevue because both my father and brother worked there.

I went to night school and got my fireboss papers. Later, I got my pitboss papers (second class ticket). The instructor was the mine manager at the Maple Leaf, and he asked me if I would like to go over there to fireboss.

At the time, it was rumoured that the Hillcrest Mine was on its last legs and due to fold up. So I took the Maple Leaf job and worked there for 27 years.

The Maple Leaf Mine was also known as the Mohawk Mine. Eventually, the Mohawk, McGillivray and the International Mines came together under one company.

When the Mohawk closed, I went to Coleman, where they asked me to be the safety engineer. That job lasted 10 years until a recession hit the coal industry.

Ethel's father operated a small coal mine near Champion during the late '20s. On occasion, she even helped her father underground.

Addie My father came to Canada from Scotland in 1907, and settled in Lethbridge when the Galts began developing their mines. He started about five different mines, including the Federal Mine, and later went on to become general manager of Lethbridge Collieries.

I was born in Lethbridge in 1909, and just naturally fell into the coal business. Winters I spent at the University of Alberta, studying mining and geology, and summers I used to work in the mines. My first permanent job was with the Shaughnessy Mine.

From 1947 to 1957, I managed the Galt No. 8 Mine in Lethbridge. It was a difficult mine, with lots of problems due to a heavy roof and heaving floor. When you're running a mine, you work hard. We had about 400 men in that mine, and I was underground most mornings by seven o'clock. When that mine closed, I moved into the oil business and then did some consulting work, ending eventually back in the Lethbridge coalfield.

Jim I was born in Scotland in 1904, and started there in the mechanical/electrical end of coal mining. In 1930, I came to Canada.

Addie and I have kicked around for a long time. We played on the same soccer team when we were younger. In Shaughnessy, we used to live next to one another. And my first job was with Addie's father in the Shaughnessy Mine.

Later, I became the chief electrician and master mechanic of Lethbridge Collieries.

Addie Donaldson, Retired Mine Manager and Jim Webster, Retired Underground Coal Mine Electrician

John I was born in 1913, in Russian country. I came to Canada after the war. In the summer, I worked on the farm with the beets. Then, in winter, I started coal mining. Later, I worked in the Picture Butte sugar factory about three months each year.

I worked 15 years at Shaughnessy, and I worked in the Russian coal industry nine years. I worked on timbering and the conveyors. There was no machine I could not work.

Sure, I like working in the coal mines because in winter it's not too cold, and in the summer, not too hot. I liked working here because I was my own boss.

Next year, I'll have only half a garden and maybe just put corn in the other half for the pigeons. I am a farmer now — I have pigeons, 12 pigeons.

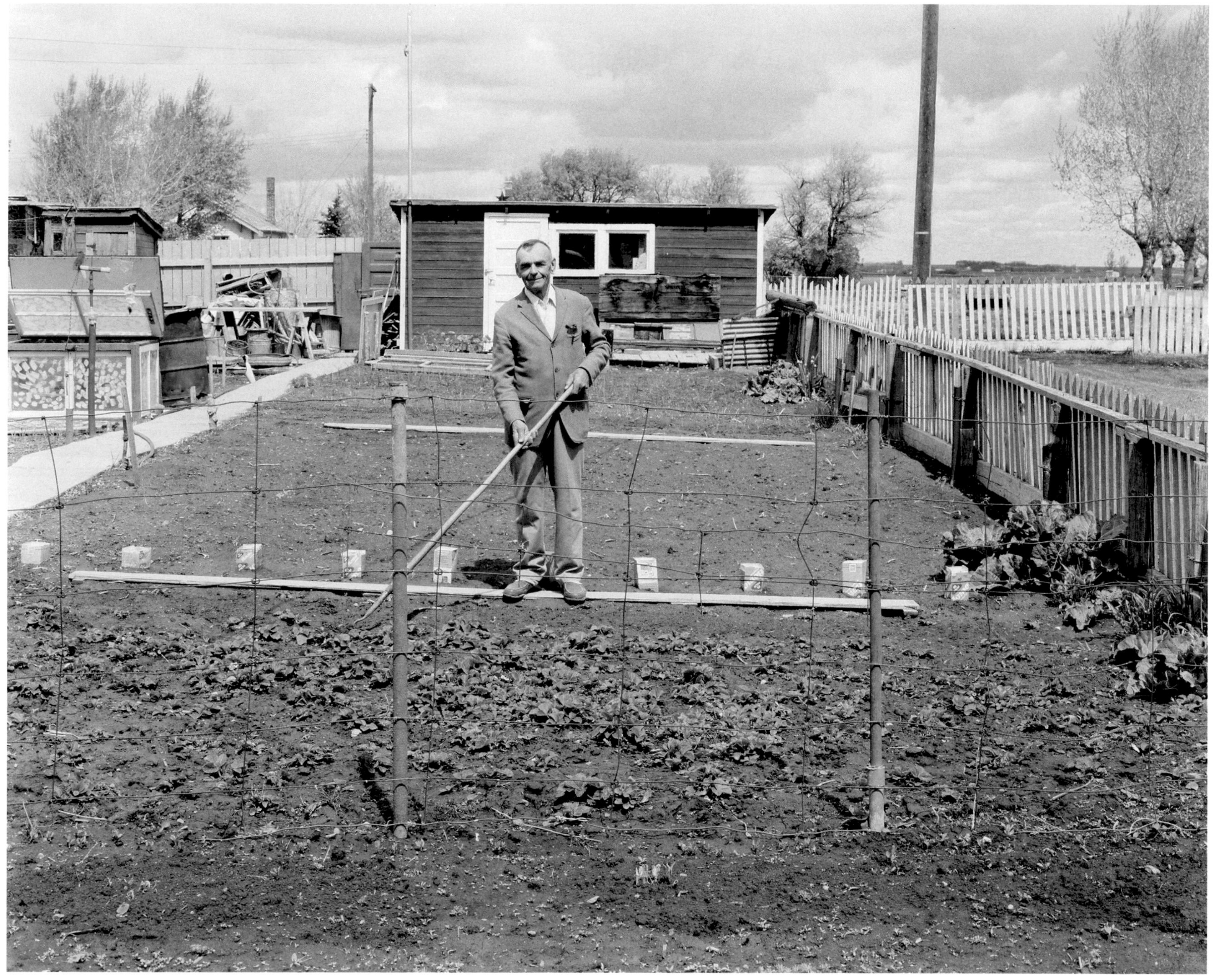

John Tuikalo, Retired Underground Coal Miner

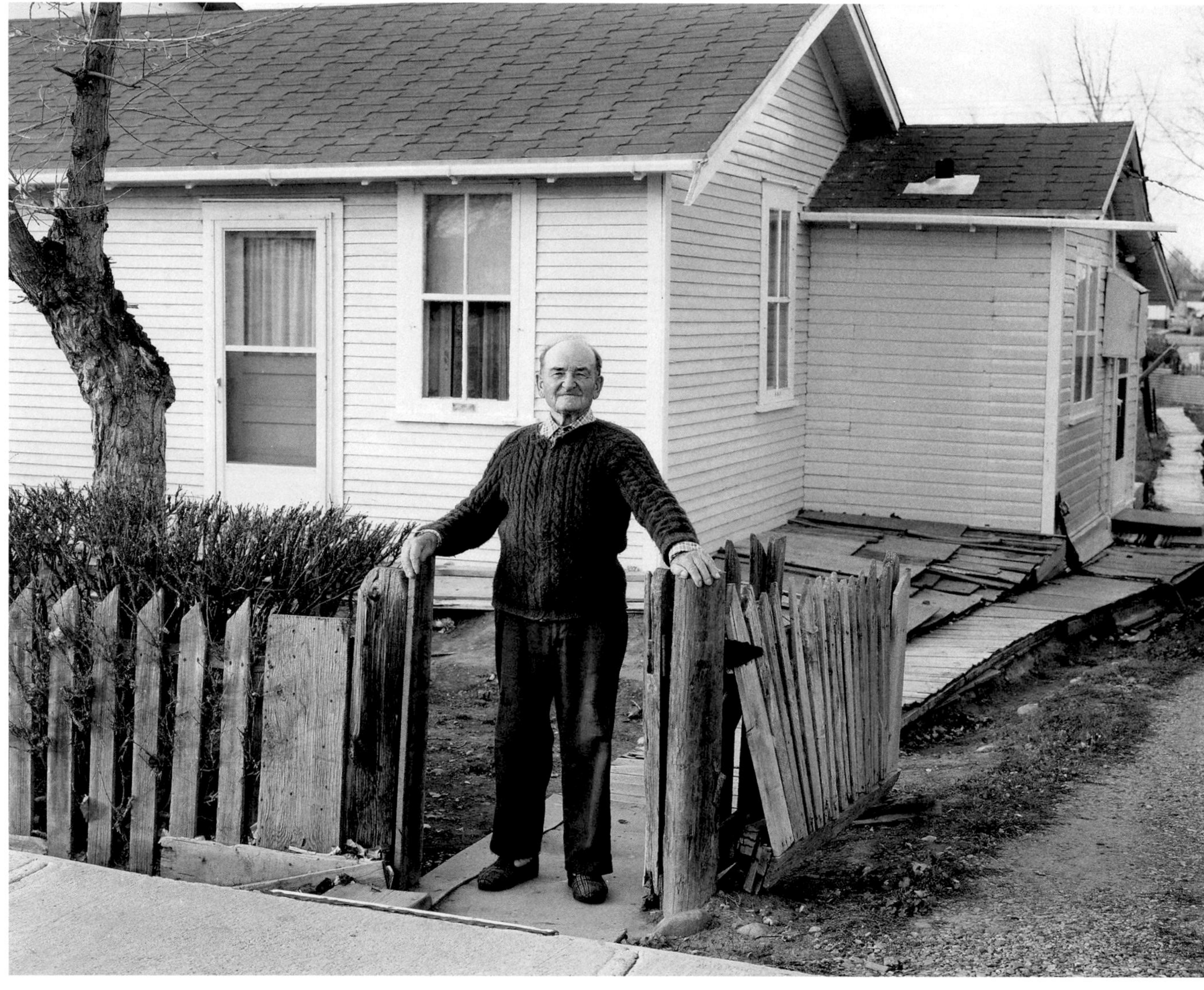

John Matty, Retired Underground Coal Miner

John I was born in Hungary in 1899. I never thought I'd live this long, because I went to war when I was 17 years old. I was a prisoner of war in Italy, and I learned to speak Italian there. I was made an interpreter, and I still speak Italian as good as anything. I came to Canada in 1925.

I worked 39 years in the coal mines in Canada. I started work at the Bellevue Mine, then at the Shaughnessy in 1928. I opened this mine up, and I shut it down, 30 years later.

Every day, I loaded 15 to 17 tonnes. I was strong, like a bull. I was kind of mad when they forced me to go on the mining machines *(other miners say John was the best machine operator, and one of the best coal miners in the Shaugnessy Mine).*

I was also an auditor of the union books, although I never did have much education. The United Mine Workers of America gave me a medal for working in the mine.

I've got to walk every day. I'm better off if I walk.

Darren Ronne, Steve Yip and Greg Hopp, Ammolite Quarrymen

Roy I was born in 1937, in McGrath, about the same time my father purchased a large farm and its mineral rights on the Milk River near Welling. As young kids, my brother and I used to find fossils along the banks of the river and in the coulees.

Ammonite is the name of the snail-like fossil, and ammolite is the name of the gemstone that comes from the quarrying of ammonite shells. For a couple of years, we called the gem Korite, after our company.

Up until my father's death in 1972, we did not recognize the commercial aspect of these fossils. We thought they were scattered all over Alberta. But as far as I know, this valley is the only place that really high-grade gem-quality ammolite has ever been found.

In the early '70s, we started to get visitors asking if they could collect the fossils on our farm. Some of these collectors were taking away truckloads of the stuff from our land, and selling it in large quantities.

So, by the late '70s, my brother Albert and I took the situation seriously, and started the first commercial ammolite mine. This quarry was situated right along side of the original homestead — subsequently, the government has declared the old foundation an historic site.

We set up a business in which we did everything, including the mining, polishing and marketing of the stone. Today, we have a contract with Korite Minerals, which is now the main operator.

As for myself, most of my life I was a farmer and rancher, but the kids do most of the work now.

Roy Kormos, Retired Ammolite Producer

Jim Belanger, Retired Clay Quarryman

Jim I retired two years ago because things were very slow. I bought a fishing boat when I retired so the family could fish in Waterton Park, and we also have a small motor home, which we take down to Yuma each winter.

I was born in Cranbrook, British Columbia, in 1926. I first started as a contractor to open up a clay pit for I-XL Bricks. I ran the operation for two years until I was hired by the company. And I worked for 37 years as part of I-XL Industries Mining Department. It was a very fine company to work for, so those 37 years went by very quickly.

A couple of my winters were spent in the laboratory learning about clays. I always had to sample and test the clays before they were quarried.

We had to move 2.5 tonnes of overburden for every tonne of clay produced. We had a couple of quarries where the strip ratio reached as high as five to one. The front-end loader was our main piece of equipment. The clay was generally found in pockets, overlaid by glacial till.

Once, we had clay pits operating in three provinces. We were supplying all the plants. We stockpiled the clays at the quarry and a contractor hauled the clay to the various plants. In the '70s, this job kept us quite busy in the field for nine or ten months of the year.

I have always been happier to be outdoors and to be working with heavy equipment. So mining was a good life for me.

From the late 1800s to 1914, Alberta had a flourishing clay and brick industry. Reserves of quality clay, water, wood and coal were plentiful in the province. And many early settlers who were experienced in the manufacture of bricks, constructed brickyards in the growing towns.

In the Medicine Hat region, the combined attributes of mineable clays, natural gas reserves and a major transportation link made the area an excellent choice for the brick and pottery industry. Redcliff Pressed Brick (later to become I-XL Industries) was established in 1912, seven miles west of Medicine Hat at Redcliff, an area with plentiful red-burning clay.

Until the 1930s, red-burning clay was mined by underground methods, with the Coal Mine Act regulating mining procedures. Today, however, the mining department of I-XL employs modern surface mining techniques, recovering clays with mobile front-end loaders.

The company's clay pits are scattered across western Canada. Red brick clays are quarried near Medicine Hat, white and buff-firing clays are stripped and stockpiled in Cypress Hills, and the Edmonton brick plant takes its clay from the Athabasca region. I-XL recovers silts from Saskatchewan, and quarries kaolinitic clays, shales, and silica for use in brick blends.

I-XL conducts geological prospecting to continually identify new sources of clay before the old deposits are depleted. To test clay deposits, analysts check drill core samples for drying, burning, colour and strength.

With its plants in Medicine Hat, Redcliff and Edmonton, I-XL is today the solitary surviving giant of Alberta's brick industry.

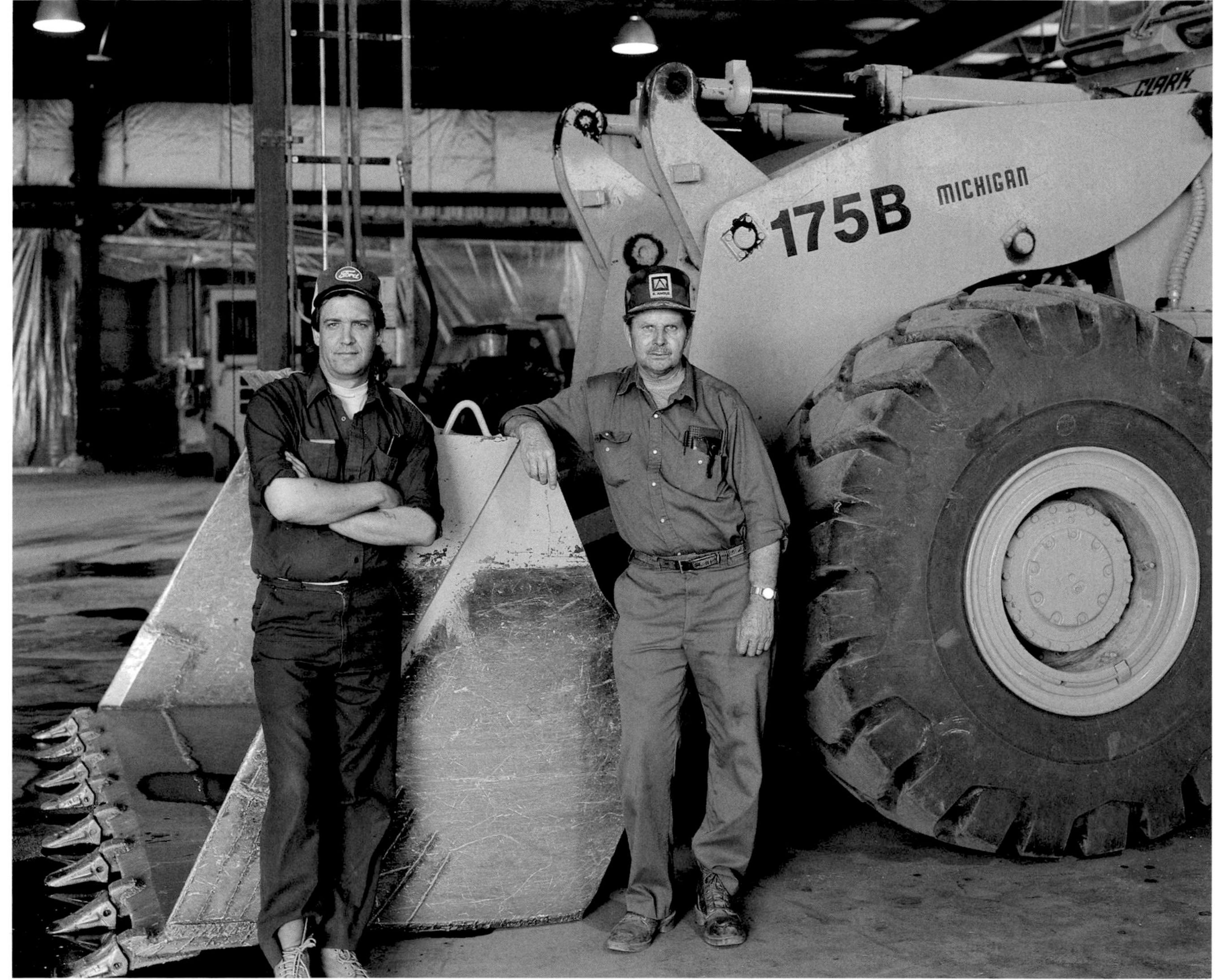

Mark Bishop and Horst Zilla, Clay Quarrymen

Mike You know, when you go in that dirty hole and you start working, you get used to it. It doesn't bother you, especially when there's machinery and you don't have to shovel. Shovelling is hard work. When you work on the pitch, you don't shovel too much. And when you come out, you forget you were working.

From 1937 to 1946, I served in the Polish army. The Russian Army made me a prisoner from 1939 to 1942 — I was awarded a variety of medals for that.

After the war, I chose to come to Canada because the climate was the same as it was in Poland. I worked two years on a farm — that was the rule — then I decided to work in a coal mine because the money was better and nobody bothers you Saturday and Sunday.

So, in 1948, I started underground coal mining as a timber packer at the Alberta Mine. Then I worked at the Ribbon Creek Mine in Kananaskis until it closed. In 1952, I started at Canmore Mines and worked as an underground miner until 1978. For 30 years, I was a coal miner in Canmore.

I've always had a garden with carrots, potatoes, cabbage, cauliflower, sunflowers, raspberries and strawberries.

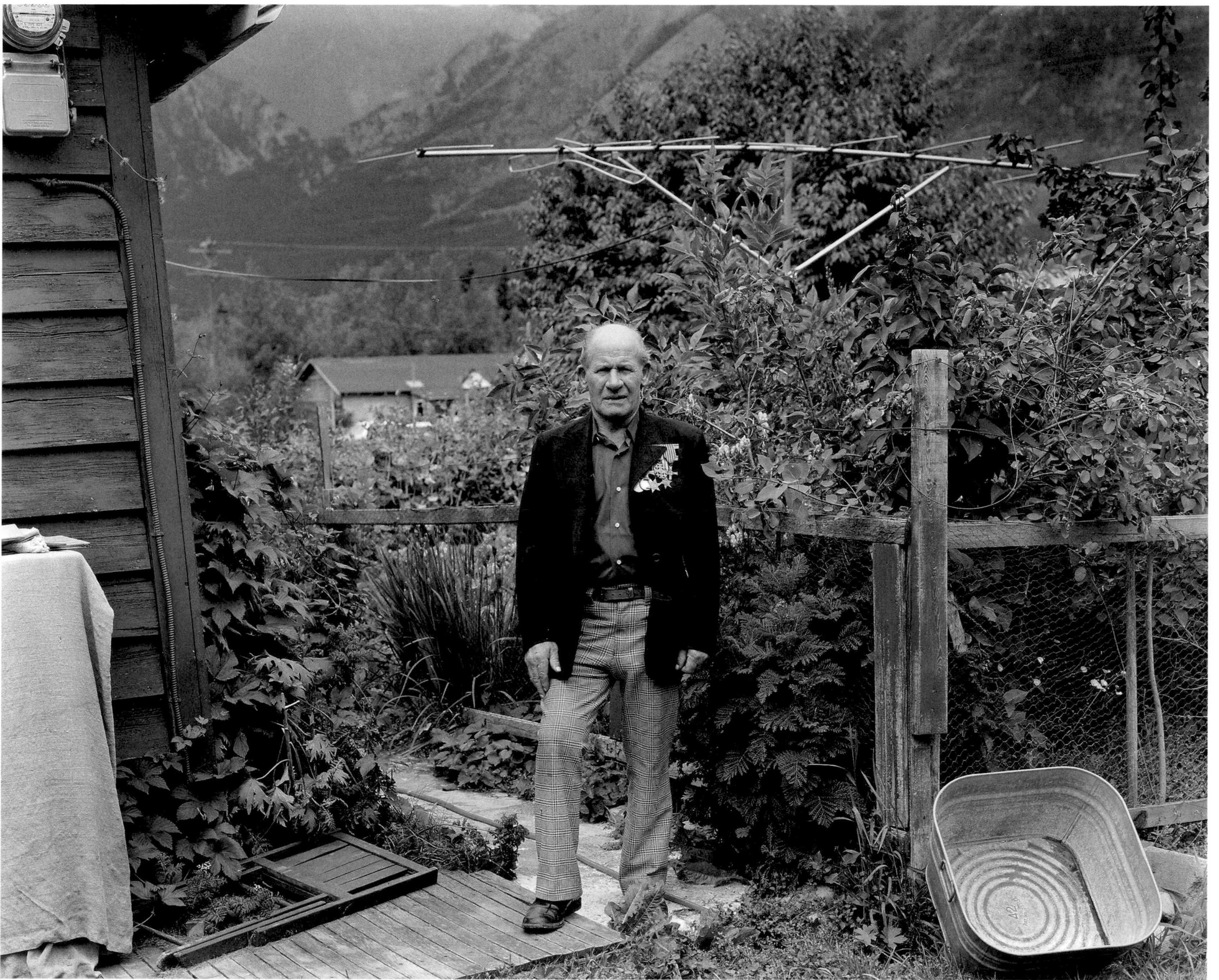

Mike Okapiec, Retired Underground Coal Miner

Jenkin

Although I've lived longer in Canada than I ever did in Wales, I am a true bloody Welshman. I was born in 1902, and went into the Welsh mines with my Dad at the age of 14. For ten years, I worked at the Baldwin's anthracite mine in Wales. Then, in 1928, we had a long strike in England. My father-in-law moved to Canmore to get established, and we followed later. But it was difficult getting a job at that point, because Canmore was only working two days a week.

When I started in the mines, all you could buy were those cloth hats. They used to get rotten with sweat. I also remember that when I started, we had candles in the mine. Later, we had the gas lamps, the carbide lamps.

It was tough during the Depression. There were few cars, so even a trip to Calgary was a big ordeal. If it wasn't for people like the manager of the company store giving you credit, half of us would have starved to death.

My daughter ran into the house one day around 1935 and said her friend had fallen off the bank into the Bow River. I ran out, saw her struggling, and dove into the river. It was winter and there was ice forming, but I was able to save her. The government gave me a medal for that.

There were only about a dozen Welshmen in Canmore at that time — true Welshmen, that is. Now we have so many different people coming into town, and they're not coal miners. They're a different class of people. They're not the same. We haven't got the same community spirit we had when the mine was going. There were about 300 coal miners in Canmore, and we were closer-knit then. In those days, if a man died in the mine, it was compulsory to go to the funeral. You could be fired by the union if you didn't go.

I got a clock for 40 years of loyal service to Canmore Mines. It's a beautiful clock, but what good is a clock when you're retired? You want to sleep in and not have to wake up.

Jenkin Evans, Retired Underground Coal Miner

John I was born in Austria in 1914, but I am Ukrainian. I came to Canada in 1929. I worked 38 years as a tipple foreman and retired in 1968, and I got a clock. I worked on the dump. I used to dump 700 to 900 cars a day. They wanted me to be tipple foreman because they could depend on me.

In the old days, everybody that worked at the mine had to walk past our house in the morning, and then back again every night when they went home.

In the old days, we had a party every week. We had a celebration. We'd bake bread. We had a gramophone playing, and we'd dance to it. Canmore was a family village. When Christmas came each year, the union and the company would donate the money for Christmas presents. We'd take all the kids to the union hall, and they'd get the presents. Today — nothing!

The town goes to hell now. There were lots of Ukrainians here, but not now. Once there were 60 men from one village living here. Now when we go for the mail or something — how many do I meet? Maybe half a dozen people who I know. All the others are strangers and the rest of them died.

In the old days, everybody was happier than now. Today, people are greedy.

This whole world has gone to hell.

Tillie and John Shachnowich, Retired Coal Tipple Foreman

Steve This mine should never have been shut down. I can tell you how much coal they left behind — it's a bloody crime and a disgrace.

My dad came here from the old country, Poland. He got crippled up at Bankhead. I was born in Bankhead in 1914, and started mining at age 15. The first day I was in there, after only two hours, they packed me out on a stretcher.

I had to quit coal mining because my lungs got too bad. I worked all my time underground in dirty, bloody jobs like rock drilling — and I got that job because I was big for my age. I got silicosis from drilling rock. I had to get out because I couldn't take the dust anymore.

In those days, you'd have a good time because Saturday you'd all chip in a dollar and buy a 16-gallon keg for $14. You'd drink for two days and go like hell.

My brothers and I all played hockey. I turned pro when I was 15 and played professionally in Seattle, Vancouver and Calgary. My brothers played in the NHL.

Everybody played sports in Canmore because nobody had a nickel to do anything else. And I'm glad I grew up during the hard times, because I had lots of sports.

But now, I bloody near die when I walk 100 feet. Yeah, with all my damn hockey medals and trophies, I've got nothing to show for it. I still have time for fishing on Lake Minnewanka with my buddies.

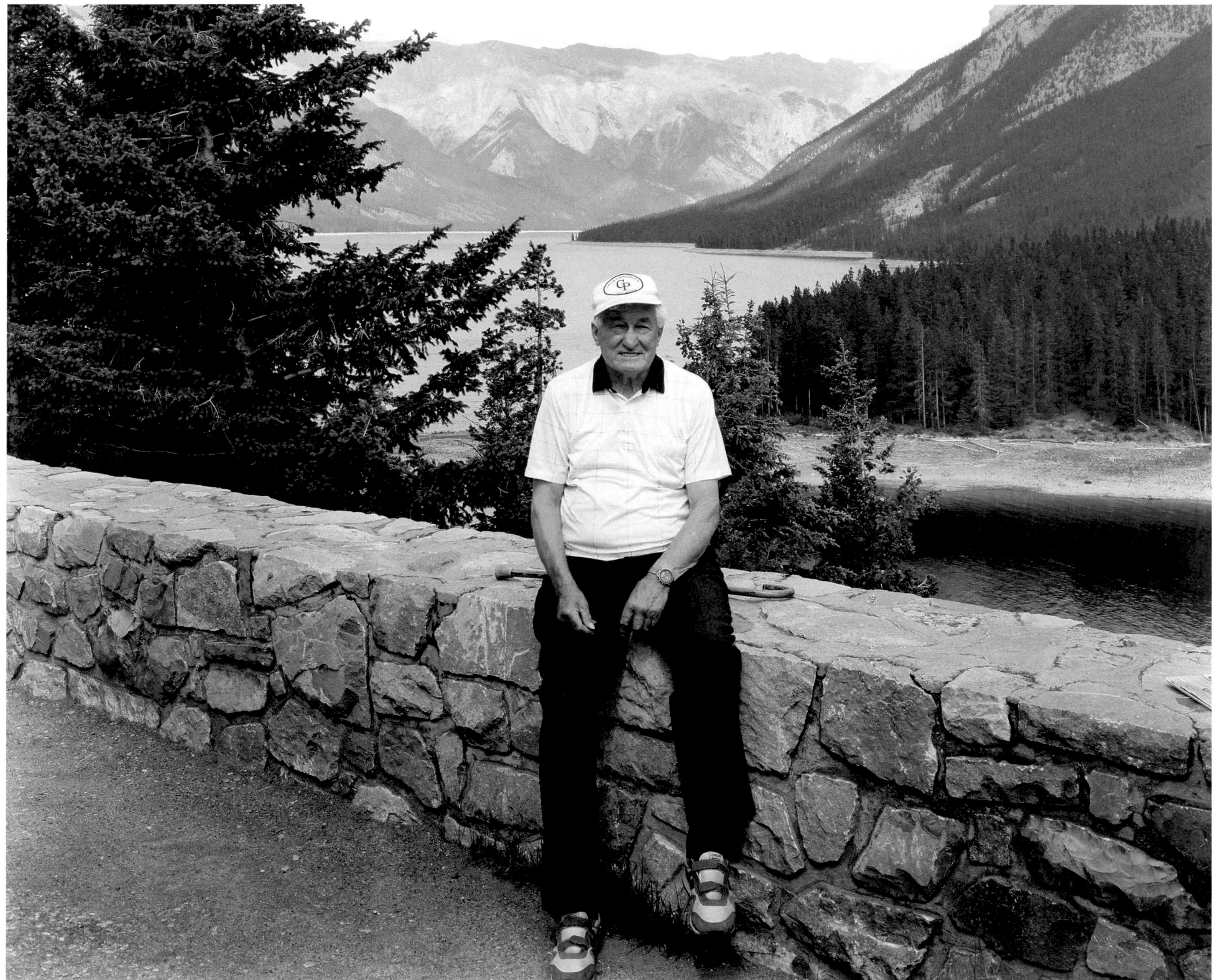

Steve Jerwa, Retired Underground Coal Miner

Roy Beiger, Retired Underground Coal Miner

Roy

I was born in Austria in 1902, in a place which later became part of Hungary. When I arrived in Canada in 1921, I started mining coal at the Midland Mine near Drumheller.

When I first came here, it wasn't easy like now. You could get no damn job. The day I applied for a job, there were about 50 men standing in line, and I was about 100 feet away from the office. The pit boss came out and said, "You come here, you big blond guy, come here." I said, "Me?" He only picked me because I was a big guy!

So I went into the office and he said, "You want a job?'" and asked me what I'd like. I said, "Dig coal," and he said, "You can't dig coal now, but you can do company work."

So he gets me to pack timber. Son-of-a-bitch, I packed timber for two years, then I went on the coal. That packing timber was harder than digging coal. And you know, when I first started mining, I had to hand-bore everything, including using a breast-auger. I tell you, that was slavery!

They laid me off because I was 65, and that's the law. But I would go tomorrow to dig coal if I could, son-of-a-bitch. I sure liked that mine! That's it. That's the life for me! I'll tell you, it didn't matter how bad it was because I wasn't afraid of danger.

All the men who worked with me are in the graveyard by the hoodoos — all except me. I was a big strong man, which is why I've lived so long. I stopped counting my age at 65, and now I've forgotten all about it....

Jim My hobby has always been collecting miners' lamps and other artifacts of coal mining. I even have my own lamp house in the basement. I've found lamps in the bush that were thrown out 50 years ago around this original lamphouse. This lamp I'm wearing is my old Wolfe safety lamp I used to wear in the mine.

I was born in Canmore in 1929. My first job was working on the picking table. A year later, when I was 18, I went underground, driving a horse at the No. 3 Mine.

My dad and I worked in the No. 3 together. One day, the mine had a big gas blowout. When it happened, I was driving the motor and my dad was mining down in No. 10 slope. I was so concerned about him that I got the pitboss, and we went back to find my dad and the others — but they'd already gone out through the top. When my dad got out and heard I'd gone back in for him, he went back to look for me. We made full circle. It was a dangerous situation because the coal started flowing and expanded to fill the whole working area.

I got my fireboss papers and was the last mine rescue training officer at Canmore Mines. At first, I was a regular team member, then I became captain, and in the end, training officer. We won the Provincial three times.

If I had to do it all over, and the mines were running like they were when we were kids, I would go back and do the same thing again. But I wouldn't want to work with the guys who were there when I retired. All those guys wanted was their paycheque. They didn't give a damn, and they couldn't work like the old miners did.

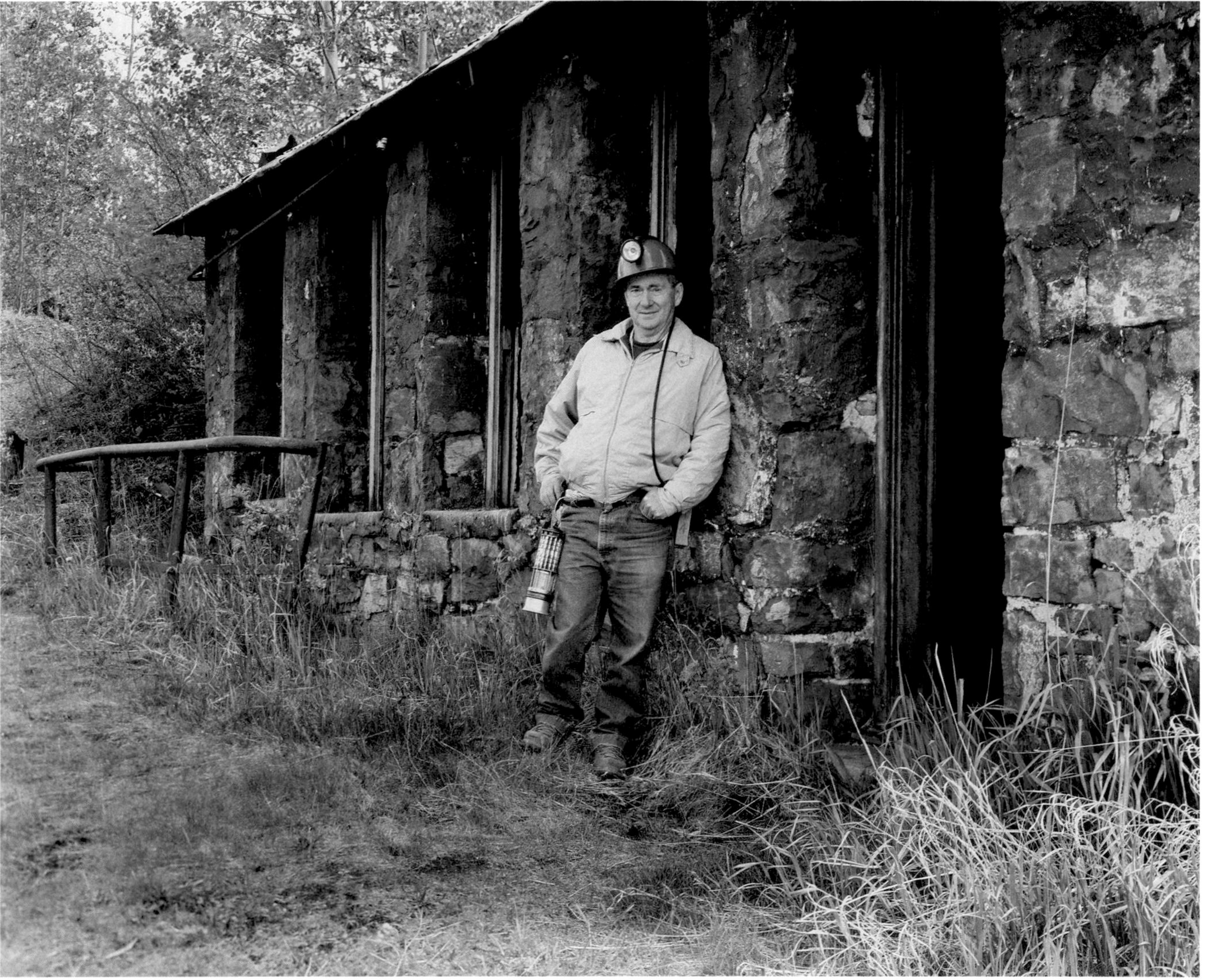

Jim Fitzgerald, Retired Underground Coal Miner

Lafarge Limestone Quarry, Exshaw

In 1906, the Western Canada Coal and Cement Company built this cement plant. Four years later, Canada Cement bought the plant and established a small village, large boarding house, a store and several company houses near the site. Since then, the plant and quarry have undergone numerous changes, with the addition of new kilns and removal of the original village to make way for extended operations.

This quarry is the largest of its kind in western Canada and produces about a million tonnes of cement annually. The site is a mountain of solid limestone, with reserves sufficient to last many years.

Lafarge Canada's cement is 80 per cent limestone, with the remaining 20 per cent consisting of black shale, which is quarried a few kilometres away at Seebee. Plant operators often add other materials such as sandstone or iron-rich additions to make different types of cement. Limestone quarrymen are concerned about stone with a high content of magnesium because it is detrimental to cement making. They selectively quarry in different spots to ensure a proper mixture of calcium and magnesium carbonates.

In the early days, quarrymen would blast huge chunks of limestone from the mountainside — but the operation wasn't a safe one. Today, the operation is a properly designed and operated quarry. Drillers, blasters, loader operators, truck drivers, cat skinners, graders, crusher operators and mechanics are the qualified workers who form the quarry operations.

Ted Fitzgerald, Loader Operator

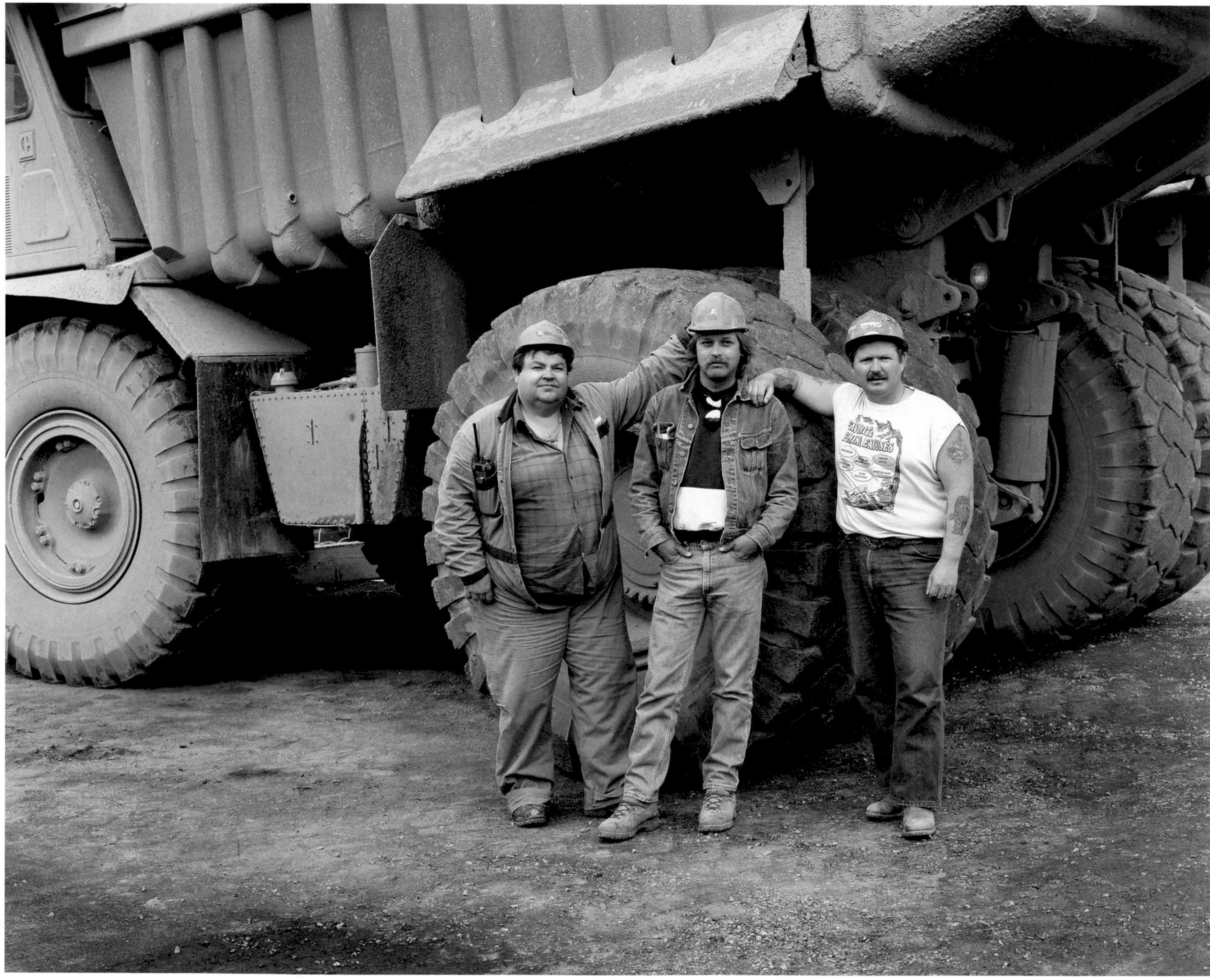

Bob Craig, Tim Ollenberger and Dan Fitzgerald, Truck Drivers

Wayne McCullough and Bernie Butkiewicz, Mechanics

On the third Thursday of each month, retired quarrymen and friends gather at the community hall in Exshaw to have their hair cut for $5 each by the "officially" retired barber, John Rock.

Back row:

Bob Hogarth
Born: 1920, Cochrane
40 years service

Denzil Jensen
Born: 1931, Calgary
40 years construction

Bob Craig
Born: 1920, Cochrane
42 years service

John Fedorus
Born: 1924, Ukraine
33 years service

Jim Bruce
Born: 1932, Scotland
Worked at University of Calgary

Front row:

Conrad Mattson
Born: 1917, Saskatchewan
43 years service

John Rock
Born: 1920, Germany
50 years Banff Barber

Ken Lyster
Born: 1915, Bomlea
31 years Exshaw Store/ Post Office

Luis Szdezo
Born: 1919, Poland
31 years service

Shorty McTaggart
Born: 1923, Northern Ireland
38 years service

Lorne Kine
Born: 1922, Provost
37 years service

Jino Lazzarotto
Born: 1924, Exshaw
45 years service

Gerry Egger
Born: 1930, Lethbridge
41 years service

Retired Limestone Quarrymen and Friends

Continental Limestone Quarry, Kananaskis

In Exshaw, the roots of the limestone industry go deep. A Scot named McCanleish was the first to burn lime in pot kilns at Kananaskis when the Canadian Pacific Railway reached the area in 1884.

By 1888, the business was doing well and supplying the thriving town of Calgary. In that same year, Edward Loder began working for McCanleish, and by 1890, Loder and his brother emerged as owners and operators of the site.

In 1905, they incorporated under the name Loders Lime Co. and built three new vertical kilns, which began burning lime in March 1908. With some modifications, this equipment operated until late 1967, when a new kiln went on-stream at the company.

In 1952, Steel Brothers purchased the plant and replaced the original vertical machine with a 200-tonne/day rotary kiln. Sometime later, the company changed its name to Continental Lime.

The limestone quarry at Gap Mountain overlooks the town of Canmore. Old-time drillers quarried the rock by hanging on ropes over the steep rock faces, dislodging the rock with hand drills and blasting methods. Later, large rotary drills were lowered from the top down the rock faces.

Today, drillers work on flat benches and operate high-speed drills capable of drilling many holes in one shift. It is not uncommon for Continental drillers to encounter curious mountain sheep who will venture as close as 10 feet away from the workers, even while the drills are operating.

Near the Continental Limestone Quarry are the remains of an old kiln and quarry. This site was started by a man called Butchart — the same Butchart who later left Alberta to open the Butchart Quarry where Victoria's Butchart Gardens are located.

David Rea, Quarry Driller

George Meier, Quarry Driller

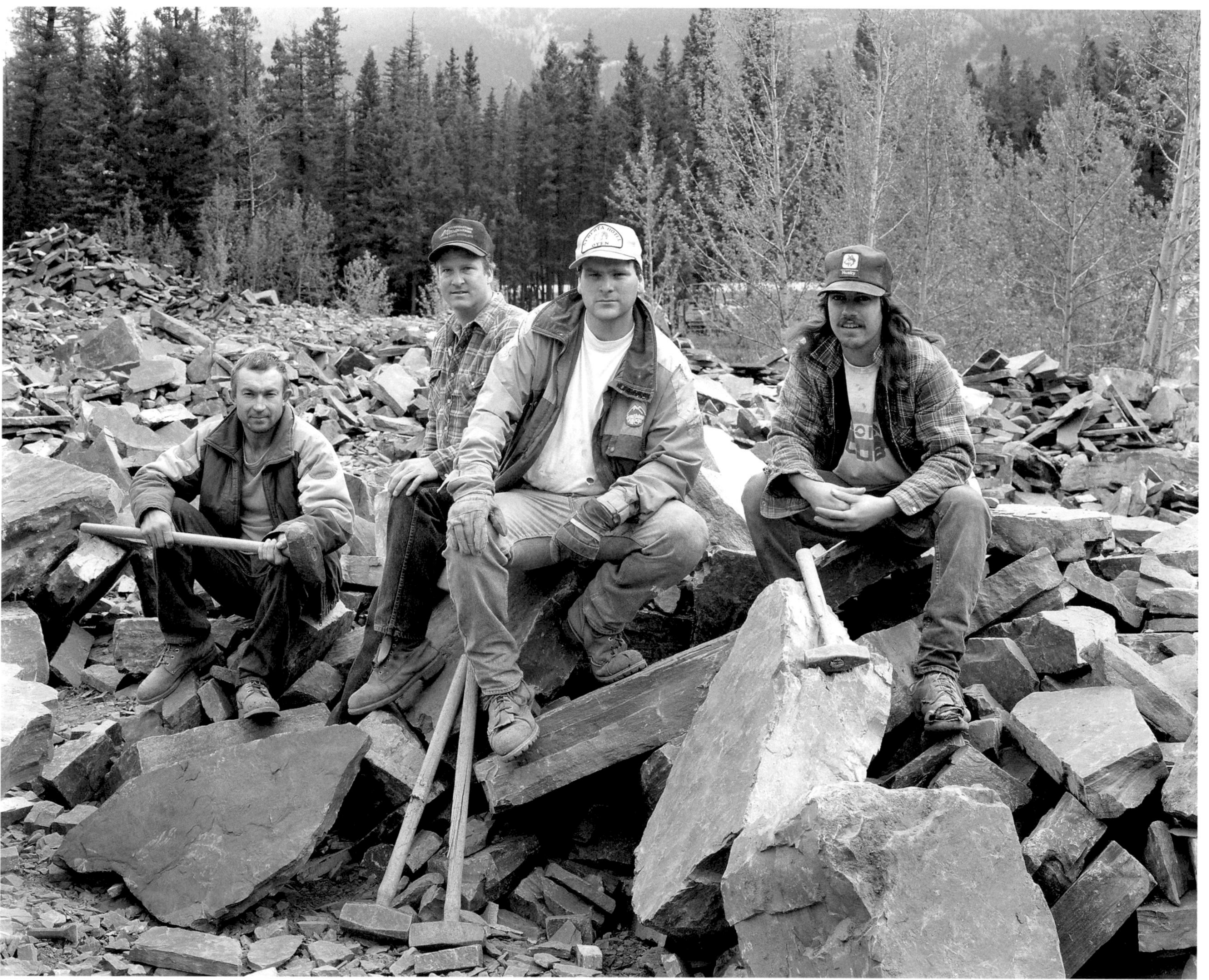

Caba Bozak, George Biggy, Mike Jones, and Evan Chambers, Quarrymen

George My Ukrainian father landed in an internment camp near Banff, and after he got out, he worked for Canmore Mines. He died of a heart attack in the mine. I tried working in the coal mine, but my father kept telling the company not to hire me. So most of my 40 years of working life was spent doing maintenance for the Highways Department.

Then, in the late '50s, the Rundle Rock became popular here. I saw that it was a good business, so I went prospecting along Pigeon Creek and found the rock lying all over the place. A couple of partners and I hired a lawyer and got a surface lease. Then we went down to the hotel and started bragging that we had the best quarry in the valley. A couple of old guys at the next table heard this. They went into Calgary and slapped a mineral lease for quarrying on the land, and then the government cancelled our surface lease. I never thought any more of it until I heard Canmore Mines was giving up its leases for Rundle Rock at Dead Man's Flats. My brother and I picked up the leases in 1961, and it felt almost like finding a gold mine.

Back then we quarried our first rock by hand. I've passed on certain techniques to my son, such as how to split the rock along the grain with stone hammers. Now the younger fellows use machinery for everything except for splitting the rock. When we started, even blasting was terrible. Today, with primacord, blasting is no big deal.

For my son's sake, I'm happy I kept at the quarry. He's made an industry out of it. George Jr. started going to the quarry when he was a kid. Today, he's 31 years old, and owns it 100 per cent. Rock has got into his blood, just like it did into mine.

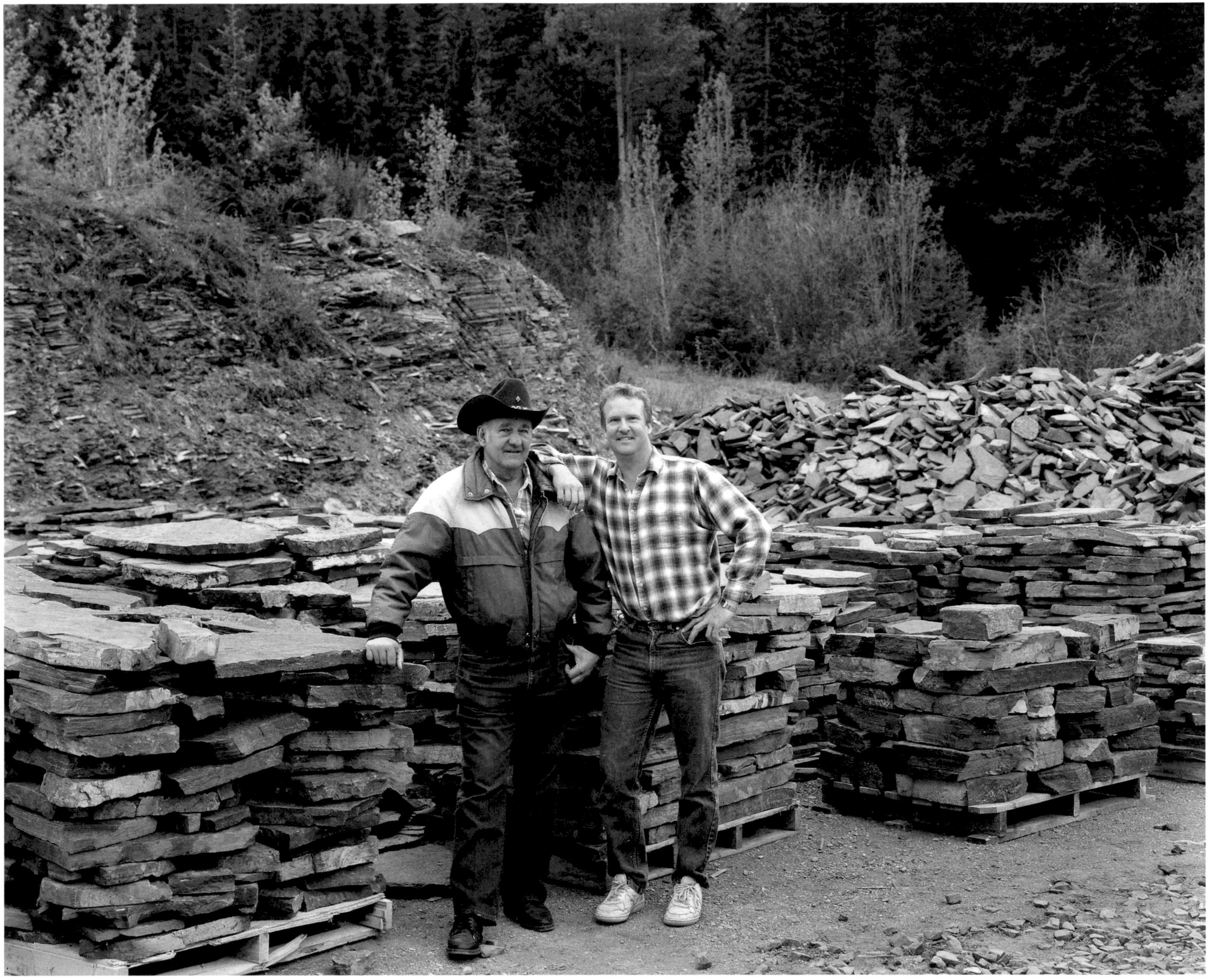

George Biggy Sr. and Jr., Operators

Thunderstone Quarry, Dead Man's Flats

Frank Marks, Truck Driver

Auntie Jude Peyto, Truck Driver

Rundle Rock Quarry, Harvie Heights

Rundle Rock got its name in the 1880s, when Canadian Pacific established the decorative rock's first quarry, on Rundle Mountain in Banff.

Many of Banff's old buildings used Rundle Rock for construction stone: the Banff Springs Hotel, the Park Headquarters and other fine buildings in town still feature the layered stonework.

Quarrying for Rundle Rock in Banff ended in the early '60s.

In 1955, Louis Kominka opened Canmore's first Rundle Rock quarry. Although Kominka was at that time a blacksmith for Canmore Mines, he was also a rock hound. During his travels, he discovered Rundle Rock in Harvie Heights, and opened his quarry even before a road was built to the site. The quarry's first Rundle Rock was hauled out of the mine in pack sacks for a Canmore fireplace.

After 45 years of service with Canmore Mines, Kominka finally retired and ran his Rundle Rock quarry full time. The business began to grow in 1980, when Kominka Jr., a geologist, joined the quarry and expanded its markets and operations. Today, Rundle Rock is used in a variety of creative ways for landscaping and for building facades, fireplaces, retaining walls and steps.

Quarrying Rundle Rock involves a series of steps. First, trees and overburden are removed, then contractors are brought in to drill and blast the rock loose. Quarrymen then hand-split the rock into flat sheets of various thicknesses. Finally, independent truckers haul the rock to individual customers.

With a growing trend to using natural stone in construction and landscaping, the future looks increasingly bright for Rundle Rock.

Dave LaFortune, Quarryman

Rundle Rock Quarry, Harvie Heights

Drew Irwin, Driller and Clay Timberg, Blaster

BURNCO Rock Products has been mining, crushing and washing aggregates from 26 pits and 18 active operations around Calgary for more than 80 years. Four generations of the Burns family have been involved in this family-owned business.

One of BURNCO's long term sites is the Springbank Operation west of Calgary, where a variety of facilities have operated since 1969. The Springbank Aggregate Pit has long been a main source of sand and gravel for Calgary's southwest quadrant, and the area could see a further 20 to 30 years of mining activity before its resource is depleted.

BURNCO's mobile mining equipment is moved from pit to pit, as required. Mining procedures involve striping the overburden above the gravel or sand deposits, then recovering the unconsolidated gravels using bulldozers or loaders. Pit-run materials are usually crushed, washed and screened to meet individual customer specifications. BURNCO carries out a continuous program of reclamation at its pits, with overburden material placed and compacted into mined-out pit areas. In the Calgary region, much of the reclaimed land is developed into residential subdivisions.

The company is its own largest aggregates customer. Uses are asphalt, ready-mix concrete and prepackaging of various production materials. BURNCO is also a major producer of shotcrete for the mining industry.

Many of BURNCO's workers have been with the company between 20 and 40 years, and overall, the operation boasts a very low employee turnover.

Max Beingessner, Foreman

Dave Shaefer, Loader Operator

David My father came to Canada from Scotland in 1911, and the rest of the family followed a year later. I was 10. At that time, there was a great emigration to Canada by coal miners, because the industry was expanding here. My father was a pit boss in Scotland when he and a number of his co-workers decided to go to Canada. He got the manager's job at the Dawson Mine in Edmonton, and he worked there until he got his papers. Then he took on the job of organizing underground mine rescue for the province.

When he quit that job, we moved to Lovett in the Coal Branch. Later, he became mine manager in Taylorton, Saskatchewan, and that's where I started underground. When he moved to Mercoal, I went along with him. I had my pit boss and manager's papers for Saskatchewan, but they weren't equivalent to Alberta's papers. So I started timbering at Mercoal, which was a new mine at the time. I ended up working as a fireboss there.

Then my dad left to become chief inspector of mines for Alberta. I worked for three more years at Mercoal, where things were pretty rough due to some fighting between unions. So I left and went to a new mine at Drinnen, three miles east of Hinton. And in the early '30s, the Mercoal finally went broke.

In 1931, my family bought the Mercoal — we borrowed money all over hell's half-acre to get that mine going again! I was mine manager as well as part owner. We lived on $125 every two months, and we were lucky. It was really tough in those days. During our first year, we worked only 50 days. There were no markets for coal.

We operated Mercoal for 10 years, and then sold it to Dunsmuir from Vancouver Island. I could write a book about the 10 years I was mine manager.

David Miller, Retired Underground Coal Miner

"Miner's life is like a sailor's
Board a ship to cross the waves,
Every day his life's in danger
Still he ventures being brave.
Watch the rocks, they're falling daily,
Careless miners always fail;
Keep your hand upon your wages,
And your eye upon the scale.
Union miners, stand together
Do not heed the owner's tale
Keep your hand upon your wages,
And your eyes upon the scale."

from (Ballads And Songs of the Coalfields)

Arthur I was born in Wales in 1908. My mother didn't want me to be a miner like my father and brothers in the Welsh mines, so I immigrated to Canada in 1924. Jesus Christ, that was the finest thing I ever done! The first year, though, if the ocean hadn't been there, I would have walked home.

My uncle had a little coal mine at Forestburg, and I went to work with him there, although I only weighed 98 pounds. In 1929, I moved to Drumheller and started at the ABC Mine. I got elected by my fellow miners and became a check weighman.

I'll tell you something: I joined the Communist Party of Canada in 1933 and I was a member until the recent parting of the ways. Every election that came up, they'd run me for the party because I was Welsh and had a big mouth. I never got elected because most of my support had no votes.

I became political, very political, and I fought. I tell you, I was like a greyhound in them days — all that was left of me was eyes and nose. In 1951, the miners decided to send somebody over to the Soviet Union, and they selected me.

I fought like hell for workers, and I'd still fight for them. In those days, the RCMP raided my home twice, but my Marxist books were buried. I still have them, you know.

Arthur Roberts, Retired Coal Check Weighman

Edith Reynolds

Edith I have very good memories about growing up in Exshaw. Of course, in those days you didn't expect what kids expect now. I remember when I was a child, I saw a car and thought how nice it must be to ride in an automobile....

I was born in England in 1907. My mother ran the boarding house owned by the Canada Cement Company. Percy, who would later become my husband, worked for my mother in the boarding house until he was 16. Then he went to the Georgetown Coal Mine, lied about his age, and got a job there.

I married Percy in 1926 and we moved to Rosedale, where he was cageman in the Rosedale Mine. When that mine closed during the summer, Percy would work in the cement plant in Exshaw, and we'd stay with my mother.

I worked in the plant a short time before I was married. In those days, the company packed and delivered the cement in cotton sacks. The used sacks were returned for repair and reuse. We'd sort the sacks, throw out the old ones, and put the good ones into a tumbler to get the dust out. I did some mending of sacks, and used to attach the tiebands to the tops before the sacks were filled.

Oh, living beside the kilns was dusty and dirty. And you didn't work too steady because the plant closed down often during the winter. You couldn't make a lot of money there — but then, of course, no one made very much money.

When the new plant was built, nearly all the old houses, including our church, disappeared to make way for the construction. So nothing really remains of the early days of Exshaw.

Percy Yeah, I spent about 38 years working in the coal mines, and I ended with nothing. Sometimes when I look back I think I may have wasted my life. But, oh golly, we had a good time in Rosedale. I used to play soccer, and everything. Oh, mining was interesting.

I was born in Kent, England in 1898, and my family came to Georgetown, near Canmore, in 1914. My first job, at 16, was trapping doors in the mine, then I went from that to pushing cars and driving. My stepfather was a teamster outside, and my oldest brother was a lampman. There were five men in my family who worked at the Georgetown Mine.

One of my brothers — the only smart one in the family — worked 46 years for Canmore Mines. He could do anything.

When the Georgetown Mine closed during the First World War, I got a job in Canmore and then went from there to the Rosedale Mine, in the Drumheller Valley. We were in Rosedale for nearly 40 years. My main job was caging at the bottom of the shaft. I had these 30 pushbuttons and I had to read the number on each coal car and ring it up to the weighman, every number. Mistakes could result in an awful mixup. There was a check weighman, a union man who counted the cars and the weight. You had to be smart to do that job.

She was tough in those days because the mines did not operate continuously. If we hadn't had the company store there, I don't know where we would have been.

When the Rosedale Mine shut down in the summer, I used to go to Exshaw. My stepfather was a foreman at the cement plant, and I worked in the packing end. I met my wife in Exshaw.

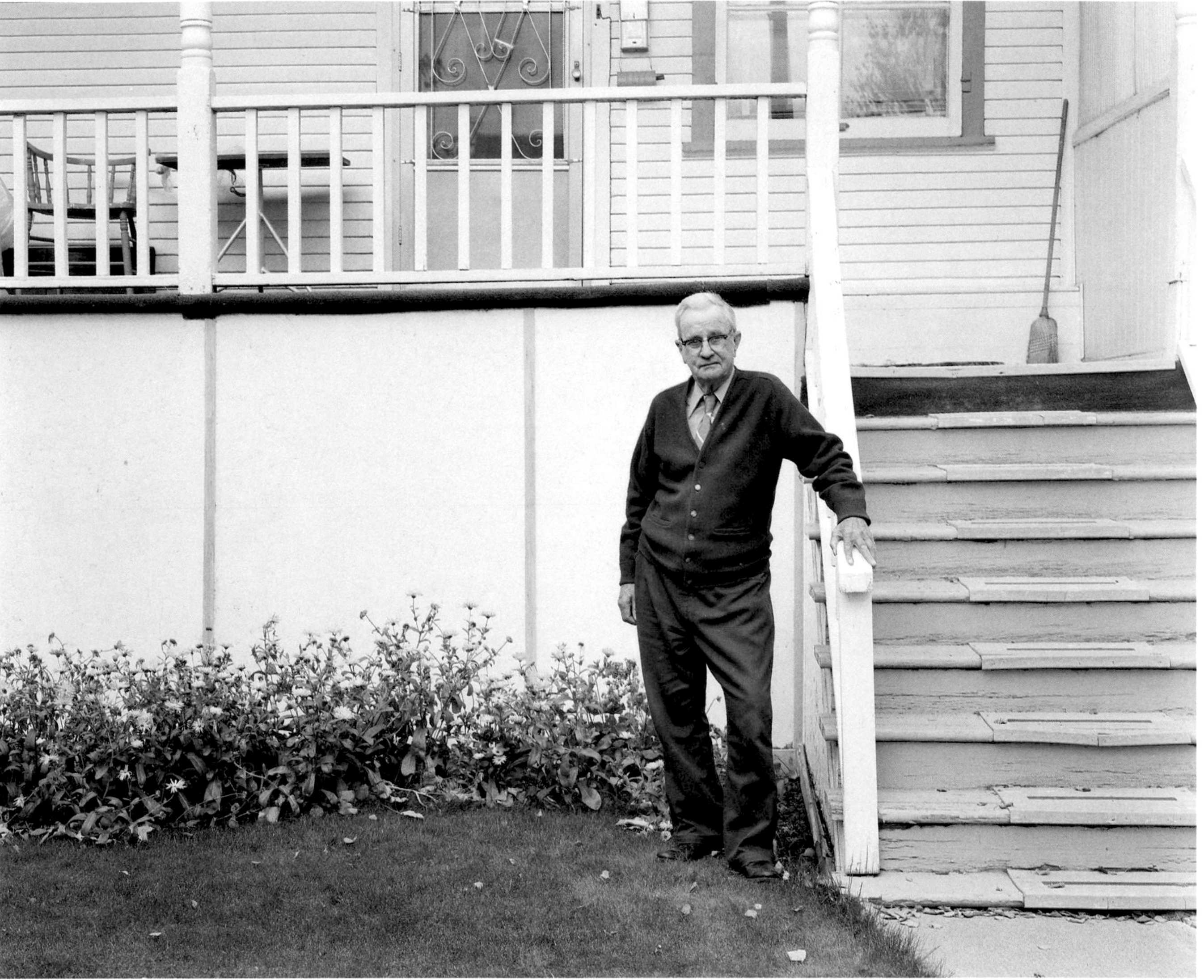

Percy Reynolds, Retired Coal Hoistman

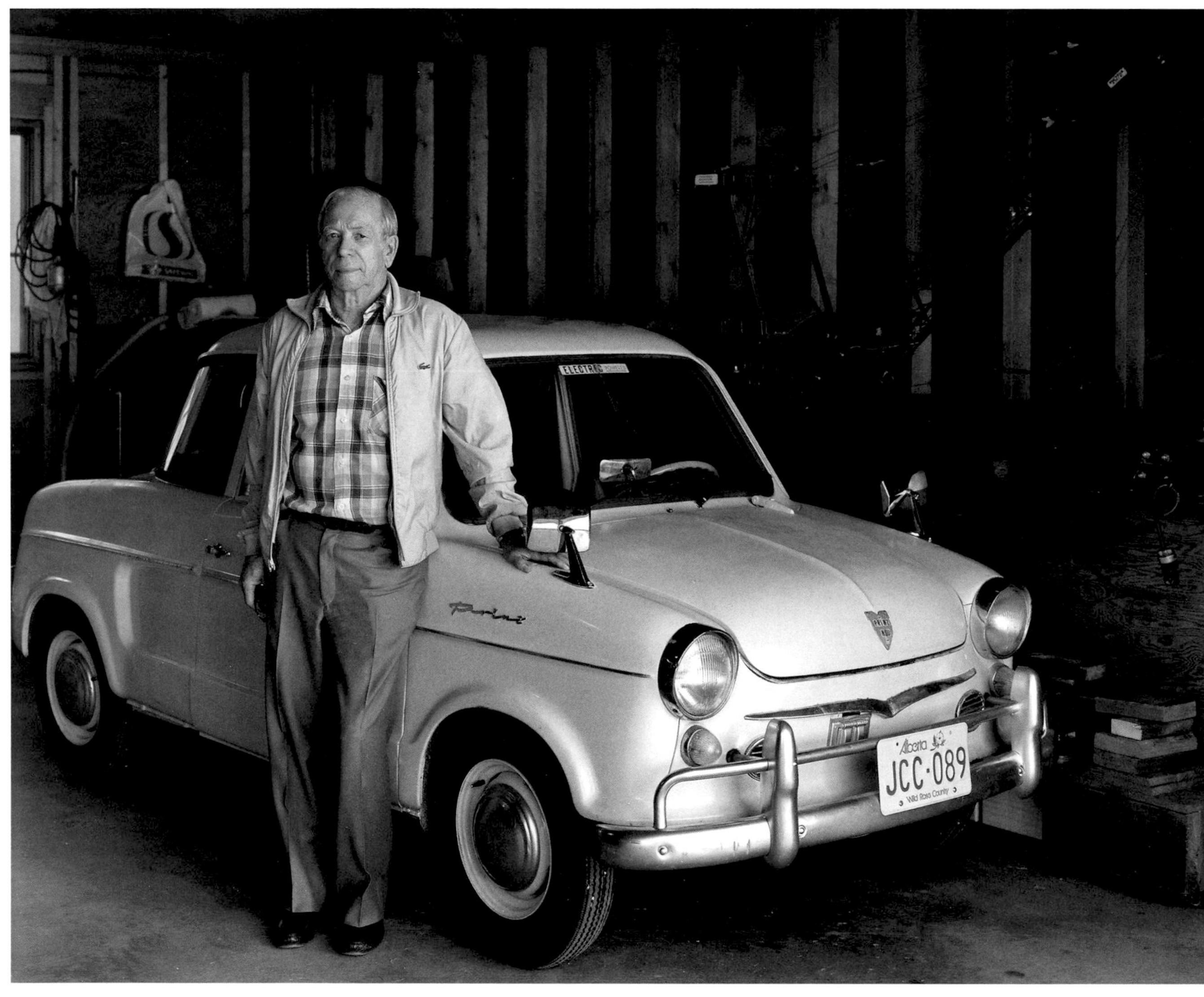

Chuck Doerr, Retired Dragline Engineer

Chuck

After the war, the most challenging thing about the coal business was finding equipment. Industry was booming and demand for coal was high. But you couldn't even buy a shovel, no way. You couldn't buy anything new, so everything you had was beat-up. That was a real challenge.

I was born in Bienfait, Saskatchewan in 1912. My schooling in Bienfait ended in grade eight, which was pretty standard because you had to go away if you wanted to go to high school. But I was always interested in electronics. By 1934, I was able to get my journeyman's papers. Then, in 1935, I started as an electrician and boilerman at Eastern Collieries. I went on to be general manager.

When it came to draglines and shovels, I made a point to understand how they worked. I'd sit up until two or three o'clock in the morning, trying to solve a problem with an electric motor. I took great satisfaction in getting a machine to produce its theoretical capacity. I learned how to draw a graph to plot these things so when I went to a shovel and dragline owner, I could intelligently say what was wrong with the equipment, and why.

Later, I was made vice president of the Great West Coal Company, and stayed until 1951. In 1955, I came to Calgary, and that's how I got on with Mannix.

In 1981, I was awarded the McPharland Memorial Medal by the Canadian Institute of Mining. The citation accompanying the award read: "In recognition of your extensive technical and practical knowledge of the coal industry and for your special skills and ability in the fields of electric power, construction maintenance and operation of large diversified electrified machines and equipment used in coal operations."

Mines/Companies

Atlas Coal Mine,
Century Coal Ltd., East Coulee

M & D Shale Quarry,
M & D Sands Trucking Ltd., Rosedale

Montgomery Coal Mine,
Manalta Coal Ltd., Sheerness

Sissons Coal Mine,
Sissons Mines Ltd., Alix

Vesta Coal Mine,
Manalta Coal Ltd., Halkirk

Paintearth Coal Mine,
Forestburg Collieries (1984) Ltd., Forestburg

Bentonite Quarry,
MI Drilling Fluids Canada, Inc., Rosalind

Dodds Coal Mine,
Dodds Coal Mining Company Ltd., Ryley

Nordegg Limestone Quarry,
Nordegg Lime Ltd., Nordegg

Coal Valley Mine,
Luscar Sterco (1977) Ltd., Coal Valley

Cadomin Limestone Quarry,
Inland Cement, Cadomin

Luscar Coal Mine,
Cardinal River Coals Ltd., Cardinal River

Gregg River Coal Mine,
Gregg River Resources Ltd., Hinton

Warburg Coal Mine,
Warburg Coal Co. Ltd., Warburg

Genesee Coal Mine,
Fording Coal Limited/Edmonton Power, Genesee

Highvale Coal Mine,
TransAlta Utilities Corporation/Manalta Coal Ltd.,
Seba Beach

Central Alberta

Locations of photographs and mines

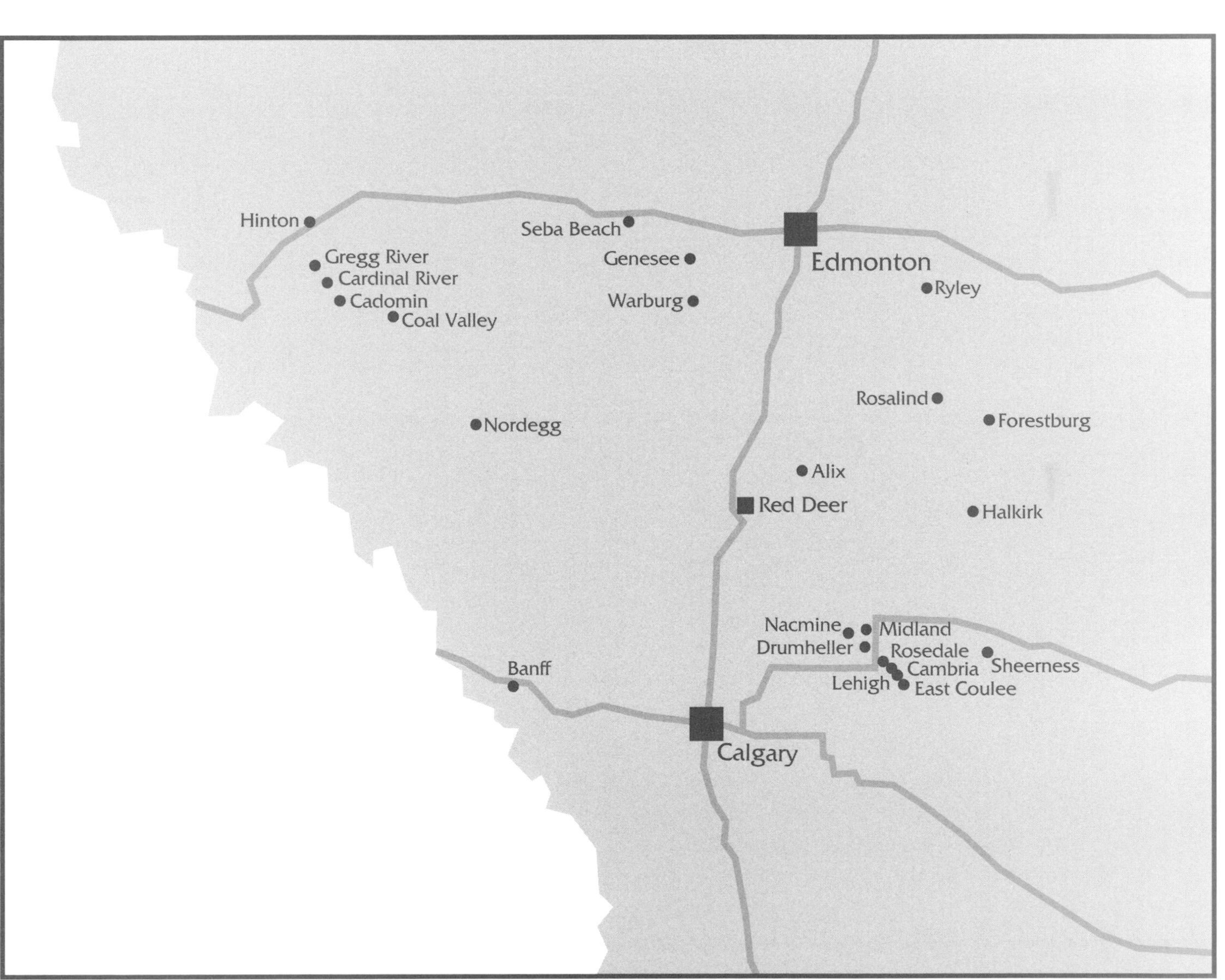

I was born in Czechoslovakia in 1902.

I first started coal mining in Wayne, with friends who were from my village in the old country. I shovelled coal into boxcars outside. There were no loaders in those days. Then I joined the army. In 1943, I came back to the coal mines in the Drumheller Valley. I went to the Empire Mine and the manager said, "I'm going to make a good driver out of you." Later, I worked at Murray, Crown and Atlas mines.

Then, in 1965, I said, "No more, I don't want to sweat no more!" I was laying track and the manager always gave me lazy men who didn't want to work. What the hell was I going to do?

When I retired, the manager called me and said, "Come, I'll give you any job, any shift you want." But no, when I retired, I didn't want to go back to work. I would have gone back to Czechoslovakia ten years ago if they'd given me my miner's pension.

Each fall, I make my own chokecherry wine. Let's have a drink — like to try it?

Martin Bystran, Retired Underground Coal Miner

Steve I am now 86 years old, the oldest man in East Coulee. You see, at my age, it could be worse. I have seen men 20 years younger in worse shape than me.

I was born in Hungary in 1900. In 1929, I left my wife in the old country with her parents, and came to Canada. It took me nine years to bring her here. In 1936, I came to East Coulee to be a coal miner. Coal mining is in my bones. I worked 27 years at the Atlas Mine and two in a gopher hole — the Etna across from the Atlas. I worked mainly as a timberman. I liked it. I was proud if I could do a perfect job.

On my last couple of years timbering, I had a partner two years older than me. He was 66 and I was 64 and I tell you honestly, no two young men put up as much timber as we old fellows. We knew every step, we never made a false move — and every move counts. Some of the younger ones just jump in there and never accomplish anything.

Many times I dream I'm still working in the coal mine, just doing my usual daily routine. I loved coal mining. In the mine, I made more money than if I was a section foreman on the railroad. I was so strong in those days I could move a mountain.

When I came home from the mine, I often went for a couple of beer — usually only two glasses. We had a hotel here that was burned down, but I don't miss it! I know my limit.

Mary and Steve Tasko, Retired Underground Timberman

Steve I was born in Hungary in 1899. At the age of 12, I started work in a Hungarian mine, loading coal.

I came to Canada — to Saskatchewan — in 1928. I worked there, digging rocks for a farmer. He paid me $40 a month and gave me board as well.

Then, in 1930 I came to East Coulee to work in the Empire Mine. Then I went to the Atlas Mine. Later, I went on to the Crown Mine, and left it in 1962. Here's how it happened.

In comes the pitboss. I want to eat my lunch, and he said, "Do you want to work again?" I said no.

He said, "I got to lay off some people — not you, just the young guys." I told him to lay me off.

"I never want to quit," I said. "I want to use the union pension."

Steve Szeman, Retired Underground Coal Miner

John I was born in Czechoslovakia in 1903, and came to this country in 1928. I arrived in Calgary, and I've stayed in Alberta ever since. I came to East Coulee in 1935, on the 20th of November. On the 21st, I started work in the old East Coulee mine, the Atlas.

I first worked as a trapper in the mine, opening and closing the doors. Then I worked for a few years outside on the picking table, then they gave me a job as a surface foreman. For that job, I loaded coal into the boxcars.

I worked from 1935 until they closed that mine down — I was 74 years old when I retired. I was the last old man to retire from the Atlas.

The company gave me a big trophy, a watch for 42 years of faithful service — something like that. They had it engraved on the back of the watch.

John Vasko, Retired Tippleman

Atlas Coal Mine, East Coulee

Julie Auld, Office Supervisor

Julie I was born in Estevan in 1919. You know, I feel good except my arthritis is bad sometimes. And I can still get around, so I want to stay here in East Coulee.

My dad was a coal miner in winter and a farmer in the summer. I met my husband, Toby, at a dance in Bienfait. He was a coal miner. We got married in Estevan in 1942. The underground coal mines were closing in Saskatchewan, but Toby wanted to continue as a coal miner so we came here. We didn't plan to stay long — just save our money and move out to the coast.

In town, I worked in the lumber yard, then at a big general store on the corner. In 1972, I started with the Atlas Mine — my first job was weighing trucks. In the end, I did the shipping, the payroll and the books, and I sold the coal. I worked at the Atlas Mine until 1987, when the owner, Omer Patrick, donated the tipple, associated buildings and the land to the Drumheller Valley Heritage Society.

I think if the mine was still operating on a small scale, I'd be hobbling over there yet. There were a lot of nice people, and I really did enjoy the work.

Steve When they started to fill the shaft at the Atlas, I did feel sad. There are millions of tonnes of coal left, but no market — if Ontario would only quit buying Pennsylvania coal...!

What kills western Canada is freight rates. The industry has been trying for years and years for a government subsidy to ship coal down east, but it's never happened.

Yeah, the hoist room and hoist are still at the Atlas. The wash house is pretty well stripped inside, but the building is still there. They want to set up the head frame at Heritage Park in Calgary.

My father came to Canada from Hungary as a coal miner in 1926. I was born in Taber in 1929.

After working in Taber, my father went to Galt No. 6 Mine in Lethbridge, where he worked three years, then we moved to East Coulee.

I was 19 when I started on surface. In December 1948, I got my miner's papers. In 1960, I left East Coulee and went to Canmore, where I did contract work and ran continuous miners.

I worked in Canmore for four years, and then I went to Coleman Collieries. After four years at Coleman, I went to Grande Cache for three years, then back to Canmore and eventually Coleman.

I came back to East Coulee in 1978. And I'll be here to retire.

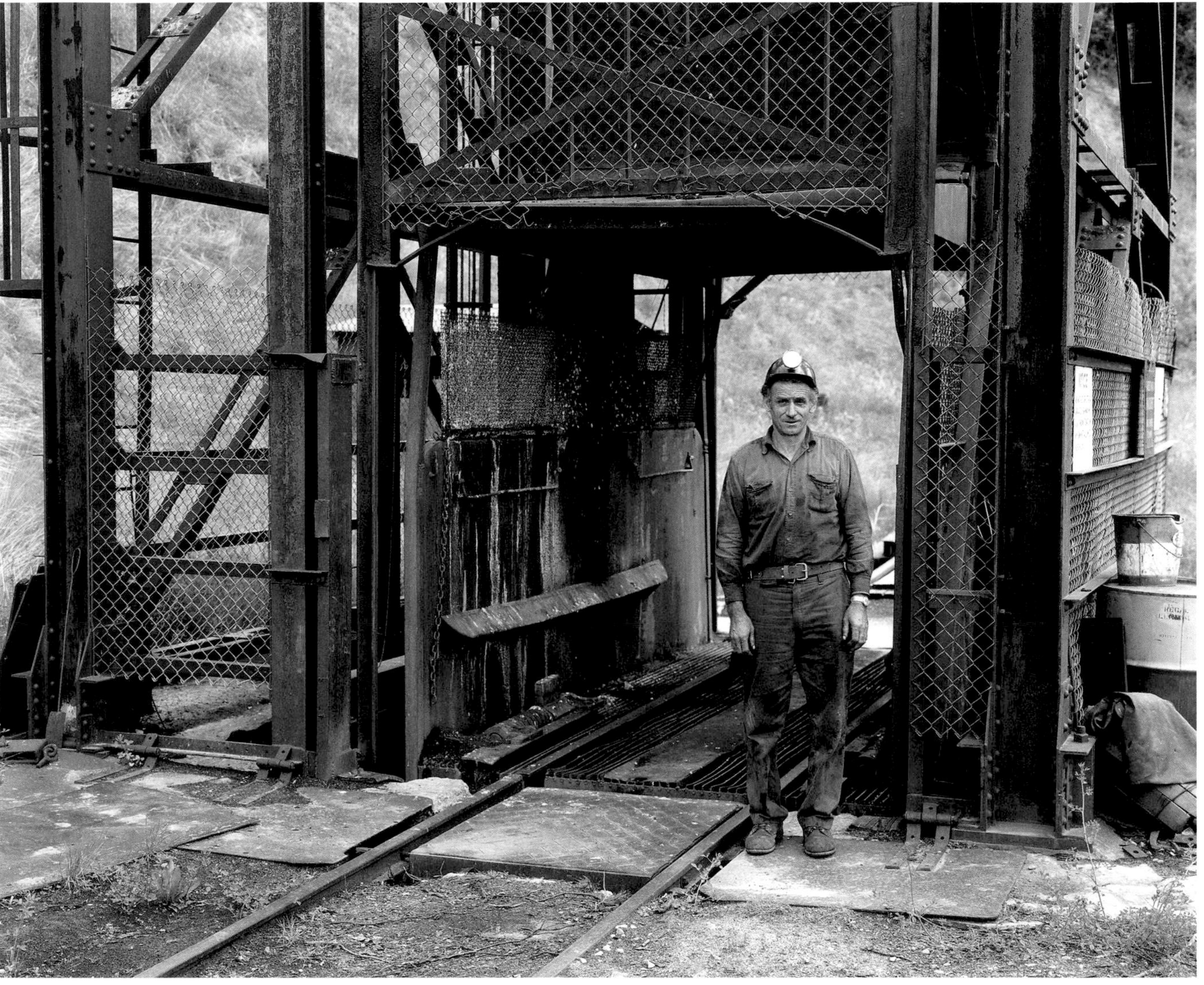

Steve Choma, Fireboss

Atlas Coal Mine, East Coulee

Operating from 1935 until its official closing in 1981, the Atlas Mine was the last large mine of the Drumheller Valley.

In its early days, Atlas miners reached the mine by an entry just above the tipple between East Coulee's two bridges. Later, the old mine was closed and replaced by the new Atlas, a few miles south of the old tipple. Miners entered the new site by a vertical shaft approximately 150 metres deep. A short surface railway hauled the coal to the tipple.

In the last few years before the mine officially closed, visitors were taken on underground tours with experienced old-time miners. The tour and accompanying commentary gave visitors a glimpse of the working conditions and methods used by East Coulee miners.

There was sincere sadness among the substantial number of Atlas retirees when, in 1984, a crew of men salvaged the mine's underground equipment and filled in the shaft.

August Gaundry
Miner, Hudson Bay, Saskatchewan

Mike Luzyzyn
Coal Miner, East Coulee

Joe Laslop
Mechanic, Drumheller

Kenny Griffen
Coal Miner, East Coulee

Steve Choma
Fireboss, East Coulee

David Daly
Coal Miner, Drumheller

Dismantling Crew

East Coulee

Situated 20 kilometres southeast of Drumheller, on the silty Red Deer River, is the coal mining town of East Coulee. In the late 1920s when the first mines here began production, the railway reached East Coulee and the town quickly grew. By the mid '30s, its character had reached maturity.

In those days, East Coulee was a one-industry town, and the viability of both town and mine depended on the success of the mine salesman, the severity of western Canada winter, and competition from other fuels. Typical of coal mining towns in western Canada, East Coulee had a cross-section of miners from Alberta, other parts of Canada, and from eastern and western Europe who came as young men in the late '20s, '30s and '40s.

Early photographs of East Coulee reveal the muddy roads, ramshackle houses, shops, hotels, and beer parlours of a town that owed its fragile existence to local coal mines. Today, however, the shiplap on the miners' houses is covered over by plastic and aluminum siding, and the families who built the town have largely passed on.

East Coulee would have become a ghost town if it were not for an influx of young people and the old-timers who remained once the mines had closed. Thanks to their faith and hard work, the town survives today.

Coal Miners' Picnic

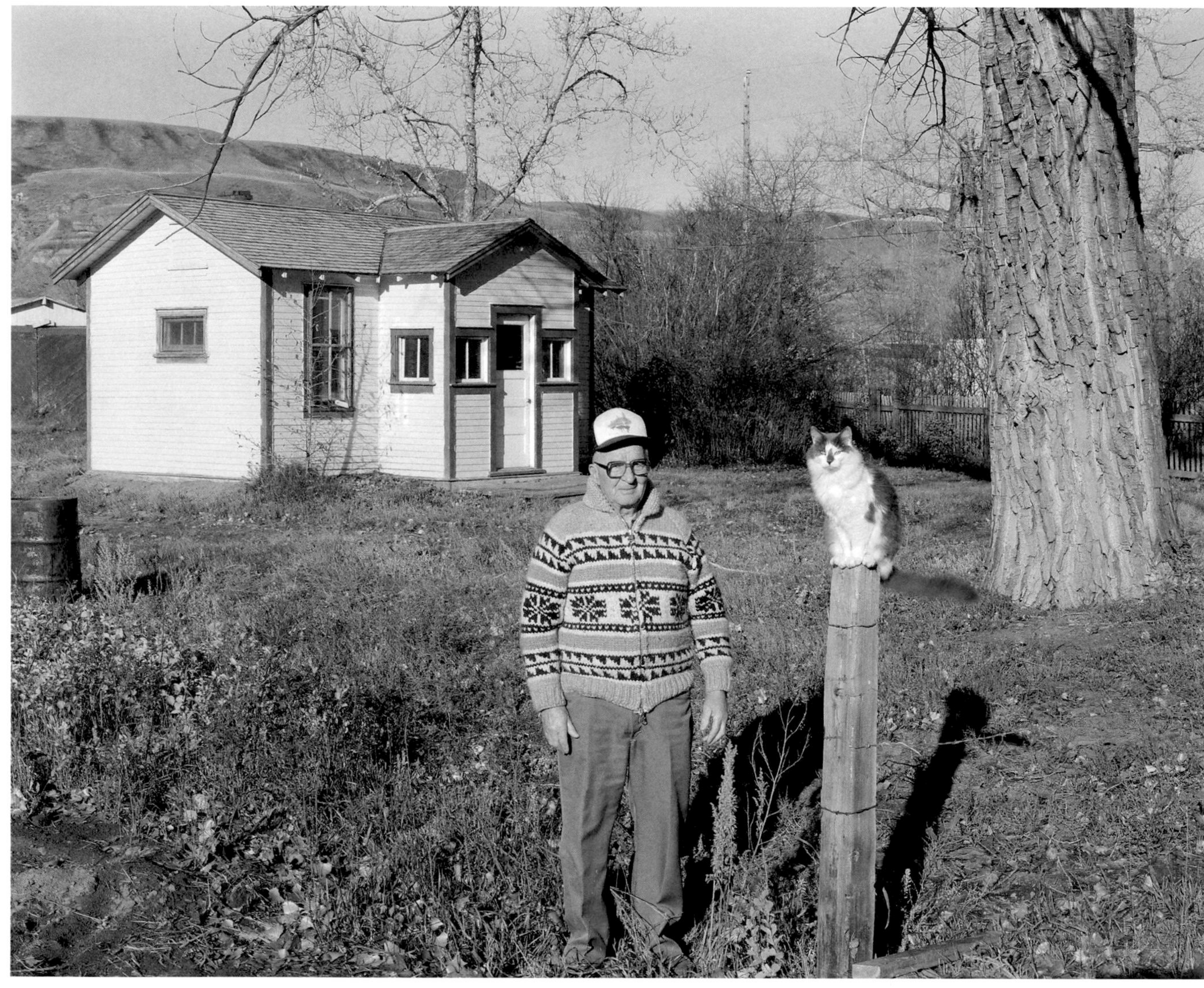

Ed Shafer, Retired Underground Coal Miner

Ed I was born on a farm at Delia in 1913.

I started in my first coal mine at Lehigh, the Maple Leaf Mine, when I was already 35 years old. It was quite a shock for me to go into a mine after farming, but I had no choice — jobs were scarce.

I moved on to other mines — the Empire and the Murray Mines — but I stuck to mining until the end. Working underground in the Empire was difficult because you couldn't straighten up when you were in there. You were lucky to be on your knees. It was hand-loading the lumps that miners got paid for...

When the Maple Leaf closed at Lehigh, I bought all the land I could get my hands on — $10 an acre. I own the little miner's shack over here too. It's original. It's been here since the Maple Leaf Mine was operating.

This farm has been my pastime. That's what I did when I had nothing else to do when I came home. You see, when I got off from the mine at 4 p.m., I used to do farming.

Now that I'm retired, I don't know what to do with myself. Even yesterday, I just couldn't hold myself down. I cut trees. I do anything I can think about.

Oh heavens yes, I'd rather be on a farm than in a coal mine, because I was brought up on a farm.

Frank I was born in 1908 in a little coal mining town in eastern Washington state. My dad was looking for a new mine to work in, and we ended up in Wayne in 1920. I've been in this valley ever since. In 1920, I had my first fight, too. I wasn't half way down the railway tracks with the suitcases and some kid challenged me...and he got it.

I went to school in Wayne until I was 15. My dad was responsible for sending me into the mines, because I wouldn't go to school anymore. He said, "Okay, I'll fix that." Well, I wasn't sorry about that because I wasn't satisfied with the teaching at the school. They were 17th century teachers, still wearing long skirts and celluloid collars.

So I hit the pits, the Jewel Collieries. I was a motorman for 20 years on the main haulage. In 1938, the Jewel Mine closed in Wayne and reopened in Cambria as the Western Gem & Jewel Collieries. When I moved here, I studied and made my third-class fireboss certificate.

I closed the mine here in Cambria in 1950. I was the boss on the job. When everything was taken out from inside, I closed up the airway and the main entrance.

After Cambria, I firebossed for another 10 years in the Atena and Murray mines. When the Murray closed in 1960, I came home and told the wife, I said, "By golly, they're closing up everywhere I go." I tried another mine, the Highgrade in Drumheller, but it quickly closed. I said, "That's it."

Cambria was filled with people when the mine was operating. When the buses went out any day, they were full. The hotel was the only place you could get beer when it was rationed. There was more beer there than you could throw a stick at.

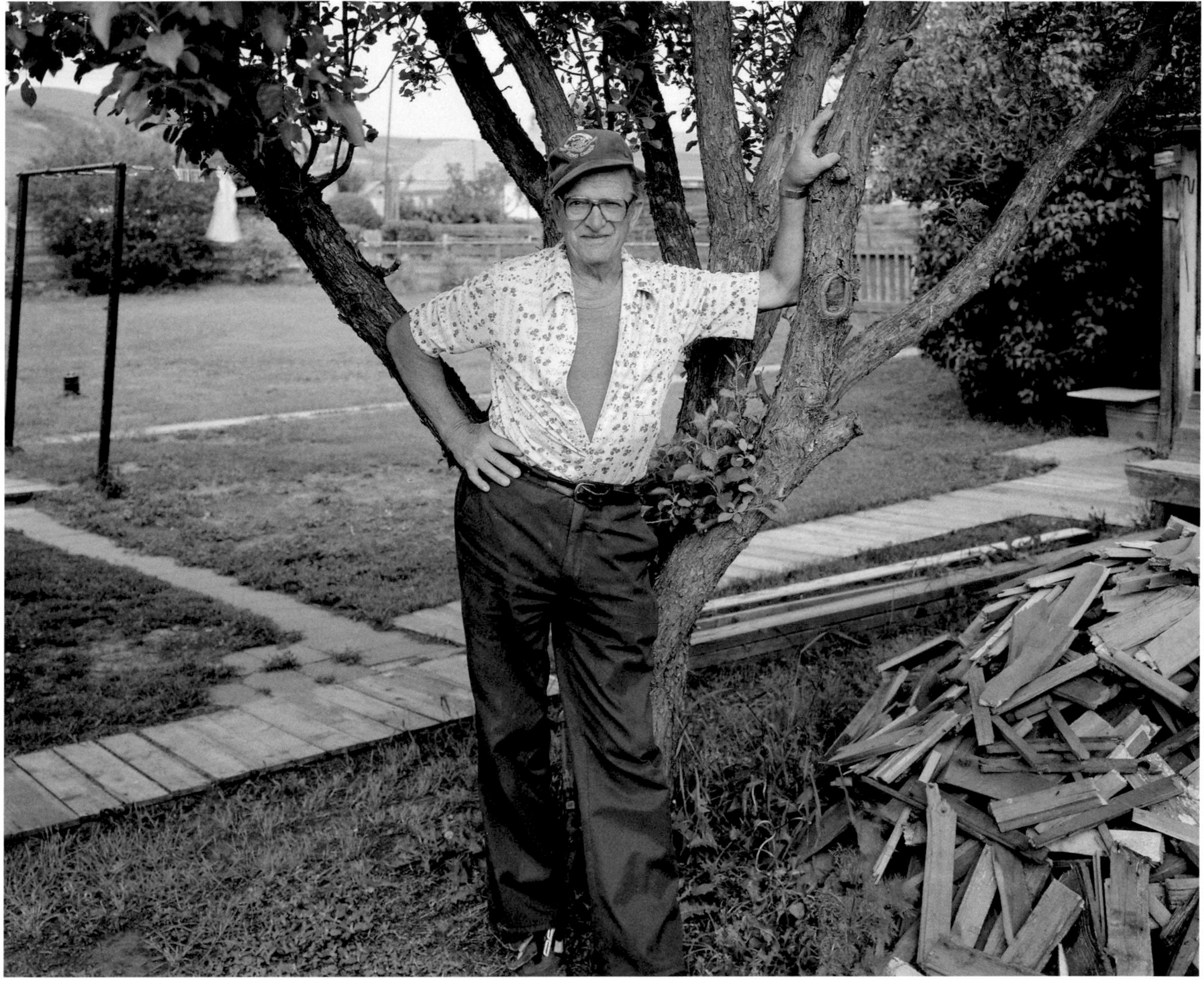

Frank Zaputil, Retired Fireboss

John I was born in Ukraine in 1900. I came to this country, not knowing a word of English. When I arrived in Canada, I went to Nordegg and got my miner's papers.

I came to the valley in 1934 and helped to organize the union. This got me listed as a trouble-maker. The mines said, "Sorry, John, we can't have you, your name is on the lists as a troublemaker." I was blacklisted for 15 months, even through a cold spell in December where the temperature reached 40 degrees below. I half starved, but there were a few faithful comrades, you understand, who worked and gave me a dollar here and a dollar there. There were days when I ate only one or two slices of bread and a black cup of coffee, but I survived. And I am not ashamed of it.

In the '40s, we organized our own little company. We all put in $3,000 each and took the land between Rosedale and Wayne — the Sunshine Mine. The mine opened in 1948. We had 36 men, and I was the mine manager. Those were six years of hard work and long hours. But as long as we had orders, it was all right and we were very successful.

After the Sunshine, I got a job at the Murray Mine and stayed there from 1954 to 1959. Then I went to the Atlas and worked to 1979. After that, I worked as a tourist guide for four years.

Also, I was a local secretary of United Mine Workers of America, and I spent 16 years as a school trustee and 18 years on the hospital board.

John Kushnir, Retired Underground Coal Miner

Vince In 1906, two years after I was born, we came to Canada from Hungary. My father worked at Michel, British Columbia for a few years, then we moved to a farm on the Alberta prairies.

I started driving horses for coal mines in Wayne, when I was 15. There is no smarter animal than a horse — once you teach it something, it never forgets. You could give me a blocky horse and I would drive it. They used to call me "Old Smoothy" with the horses.

I've seen quite a bit of tragedy in mining. Once I saw a father and son killed right in front of my eyes.

I worked 54 years in coal mining — and in the end I just hated to leave it. I was at home in the mine. It was nice and warm in the winter. You'd never catch me working outside. The only reason I ever did quit was because when I was 72 years old, I was driving a car with a neighbour and we hit a bloody train here in Rosedale. Both my legs were hurt. That's what made me quit.

Otherwise, I would have still kept going.

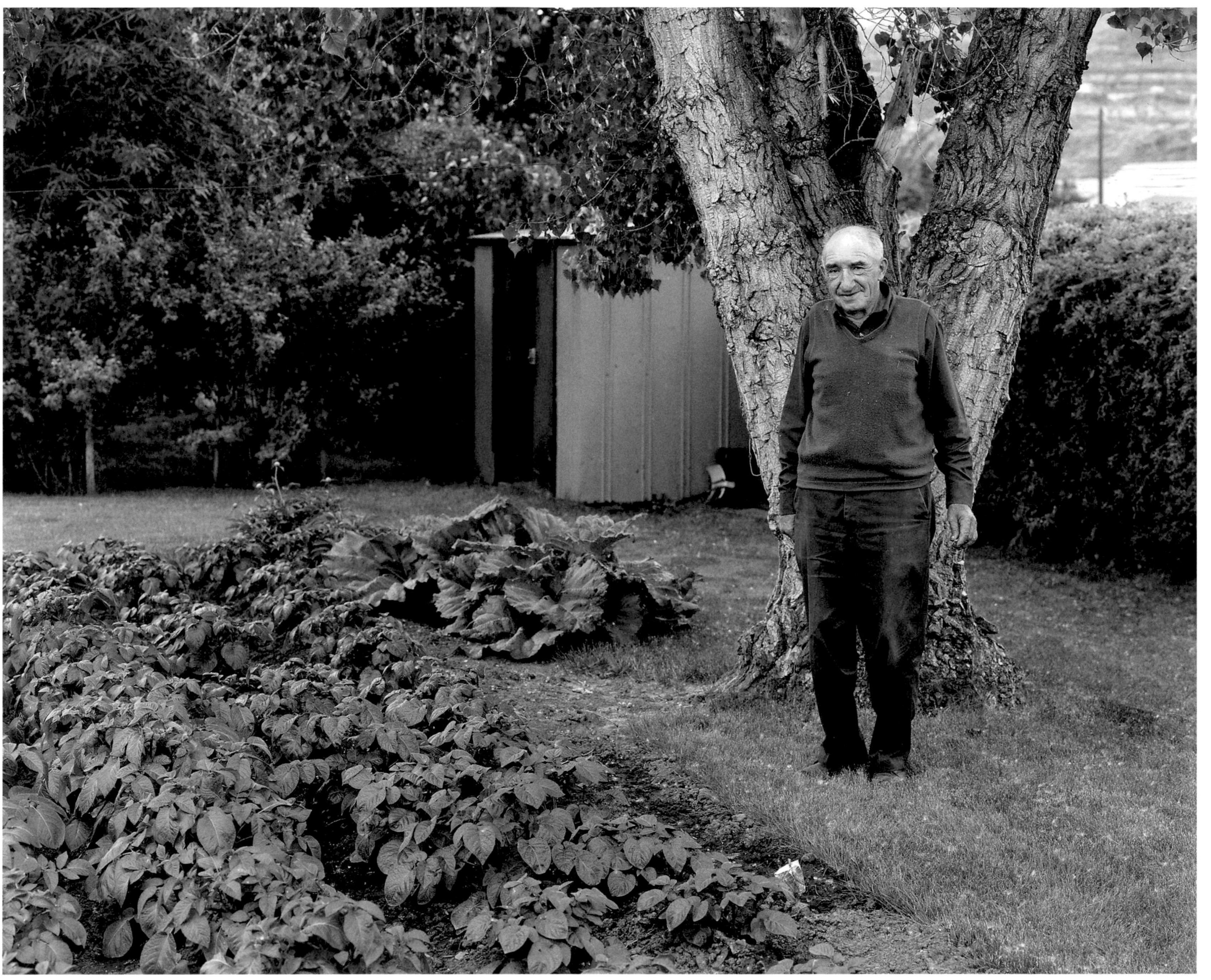

Vince Pap, Retired Underground Coal Miner

Mark The government is fooling around spending thousands of dollars on museums in the valley. Why don't they do something useful? We need bridges in this country, we need roads. All they talk in this town is tourists, tourists, tourists — it gets sickening. What does a tourist bring if he's here? July and part of August, and then he's gone home.

I was born on a farm near here in 1921. We've been quarrying red shale and strip mining coal in the Drumheller Valley for 33 years. The government boys told us we were a bunch of stupid farmers, and didn't know what we were doing. Oh yeah, that's what they told us. What did we know about mining when all the big mines were shutting down?

I used to haul waste away with my trucks for the old Rosedale Mine. Whatever I salvaged was mine to keep. I used to salvage maybe 25 tonnes a day of bone coal and grey coal. The Hutterites started buying it from me. Finally, the old mine manager came by one day and said he should have kept the dump and given me the damn mine.

When the Rosedale Mine closed, I had all these coal orders from farmers, so my brother-in-law and I decided to start strip mining just west of Cambria. We called it the Subway Mine — opened it up in 1952 and closed it in 1975. I still get people calling for coal.

There is no future left for coal in the valley. You've got too lazy a generation of people in Canada. They're not going to go underground to dig coal! You've got to bring Hungarians, Ukrainians, Poles, Czechs, Italians — they're the hard workers.

This office, built in 1912, is the oldest office in the valley. It belonged to the Newcastle Mine in Drumheller. We moved it to Lehigh where we were recovering some shale piles.

Mark Sands, Operator

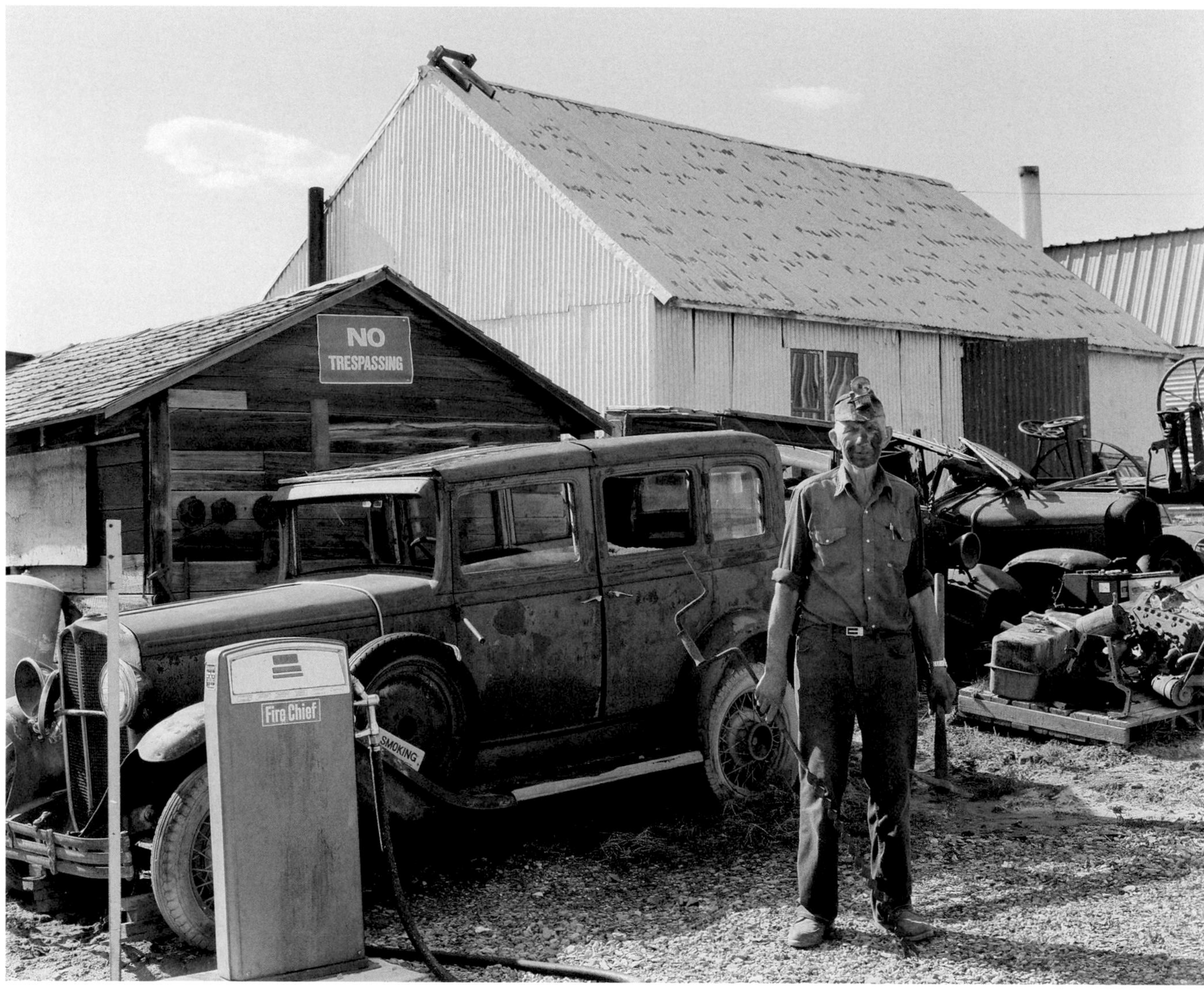

Pete When I first went into the army at Debert, Nova Scotia, they called us out on a parade and asked, "Who here is a qualified coal miner?" There were quite a few of us who stepped out. We all got a ticket home to go back to work in the mines.

I was born in 1917 in Wayne. When I was laid off at the railroad, I got a job in a little gopher-hole mine at Ward's Lane, loading coal in a boxcar. I was 17.

I ran a duckbill loader in the mine — and always on the graveyard shift. They kept switching me around from crew to crew on the graveyard. Still, I think coal mining is one of the best goddamn jobs a guy could get. I liked it.

When I saw the writing on the wall for mining, I looked around and started something else. In fact, I was still working in the mine when I started to run buses.

I've worked hard all my life. If I don't work, I'll kick the bucket. You know, I don't know how a guy can just live to sit. The only time I watch TV is to see hockey, baseball and football.

I'd like to have more time to collect fossils and restore antiques. In my spare time, I also collect stamps and polish stones for mounting in jewellery. But there's no time to do any of that.

Pete Ludwig, Retired Underground Coal Miner

Otto We all have to be good for something. I never found out what I was any good for. I followed things because I had to get a dollar on the table for grub for the family. That's all. I didn't particularly care for coal mining, but it was a job that paid fairly good money. With five kids, I needed a good paying job.

I was born in Manitoba in 1906. Originally, I moved out west to start ranching, but in 1944, I came to the Drumheller Valley to go into the mines. Still, I also managed to maintain a ranch near Fish Lake.

I started in The Empire Mine, where they made me barn boss because I was very experienced with horses. Then I started digging coal at the Arcadia Mine at Willow Creek. My last underground mining was at the Atlas Mine in East Coulee. Then they made me a security guard. In 1981, we moved from Willow Creek to Drumheller.

I am over age now and can't work like I used to. I got crippled up a time or two — that catches up with you after a while.

Otto Weisner, Retired Underground Coal Miner

John I was born in Lethbridge in 1905. My father was born in Czechoslovakia and was a coal miner in various mines in the Lethbridge and Crowsnest Pass regions.

I started mining in the Galt #6 Mine in Lethbridge when I was 16. Later, I worked in mines in the Drumheller, East Coulee, Lethbridge-Shaughnessy, Coleman and Medicine Hat areas.

I held a variety of jobs — trapper, horse driver, rope rider, mining machine operator, fireboss and finally, mine tour guide at the Atlas Mine in East Coulee.

I considered myself an expert at breaking in and driving horses. Once, I drove a three-horse spike team, hauling coal underground.

During the summers, when the mines were closed, I played professional baseball for the Galt Miners in Lethbridge and for a team in Vancouver.

I wanted to quit the mine when I was 70, but the Atlas people asked me to give tours at the site. Finally, at the age of 75 and after 58 years in the mines, I said it was time to quit.

In the summertime, you want to be outside. When that old winter comes and it starts to get cold, you head down that hole and you get used to it. I enjoyed the mines but there were some tough times.

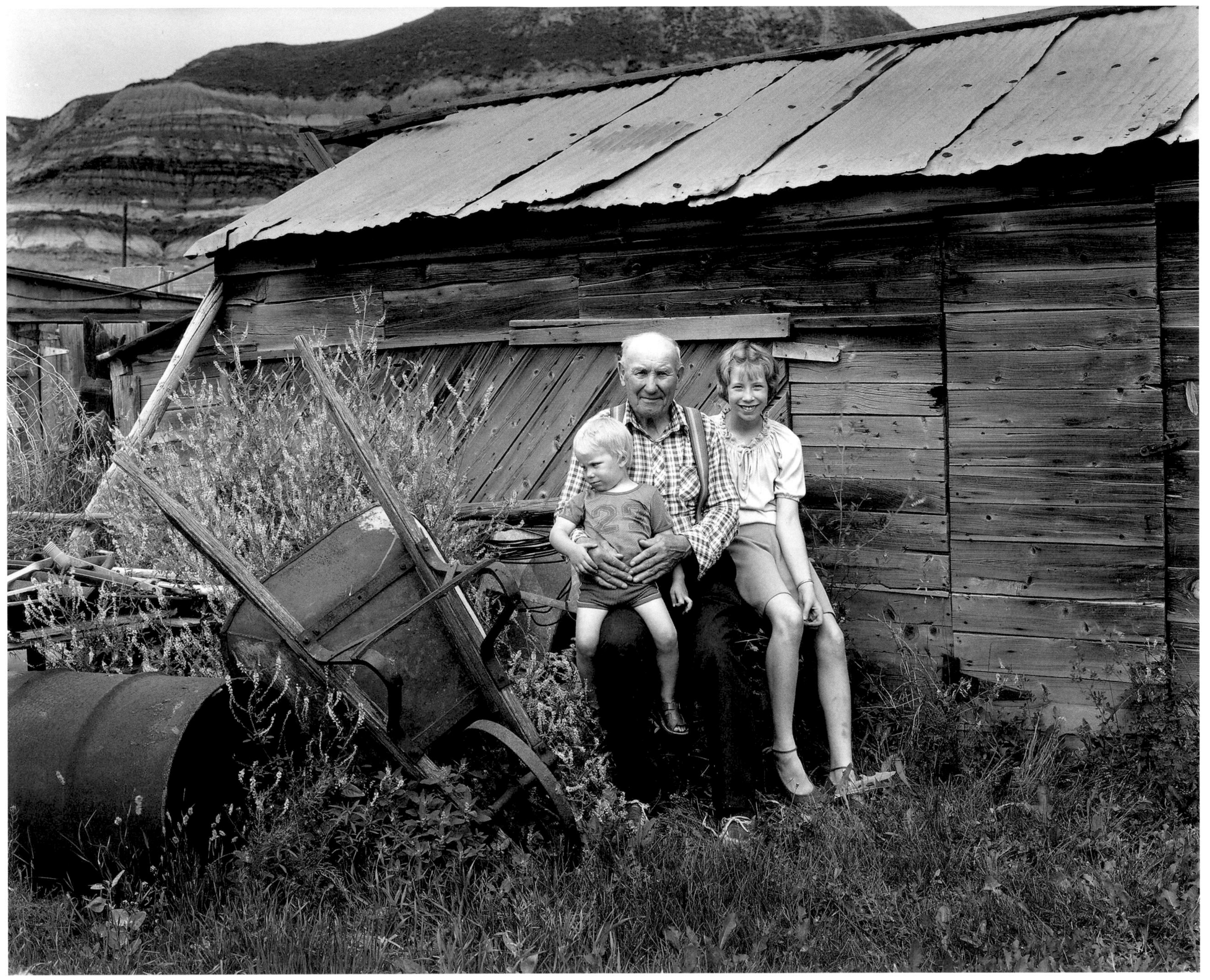

John Yorke, Retired Underground Coal Miner and Grandchildren

Coal mining was one of the best lives I could imagine. If I had my way, I'd do it all over again.

I was born in 1905, in England, and came here when I was seven. My dad started farming in the valley, and to survive, he worked in the two Midland Mines for 55 years.

I started at the Midland Mine here when I was not quite 16. At first, I helped my dad underground. Then I dug coal for 16 years. Eventually, I became head electrician at the Midland. I worked right through until the mine amalgamated with the Crown Mine in East Coulee, and then shut down. But I couldn't get along with the owners of the amalgamated mine (eventually the Atlas Mine). I had a lot to do and all they did was stand around and pat each other on the back and say what swell guys they were. I couldn't take it any more, so I left and went up to Canmore as an underground mine electrician for six years. Then, in 1970, I retired.

When I first started mining, I was in the United Mine Workers of America, and still was when I quit. In fact, I still pay my union fees of $1.50, but they give me a pension of $220 a month now.

A year ago, there was a fellow member who passed away in Manitoba. He didn't have any family or relatives at all, so he left his money to his UMWA brothers in this district. Every one of the old-age pensioners got $145 from this fellow's estate. He must have been a very good person.

It's funny: people would say to me, "You worked in the mines.... Weren't you afraid to work down there?" Well, I told them dozens of times that I felt safer in the mine than I did crossing the streets in Drumheller.

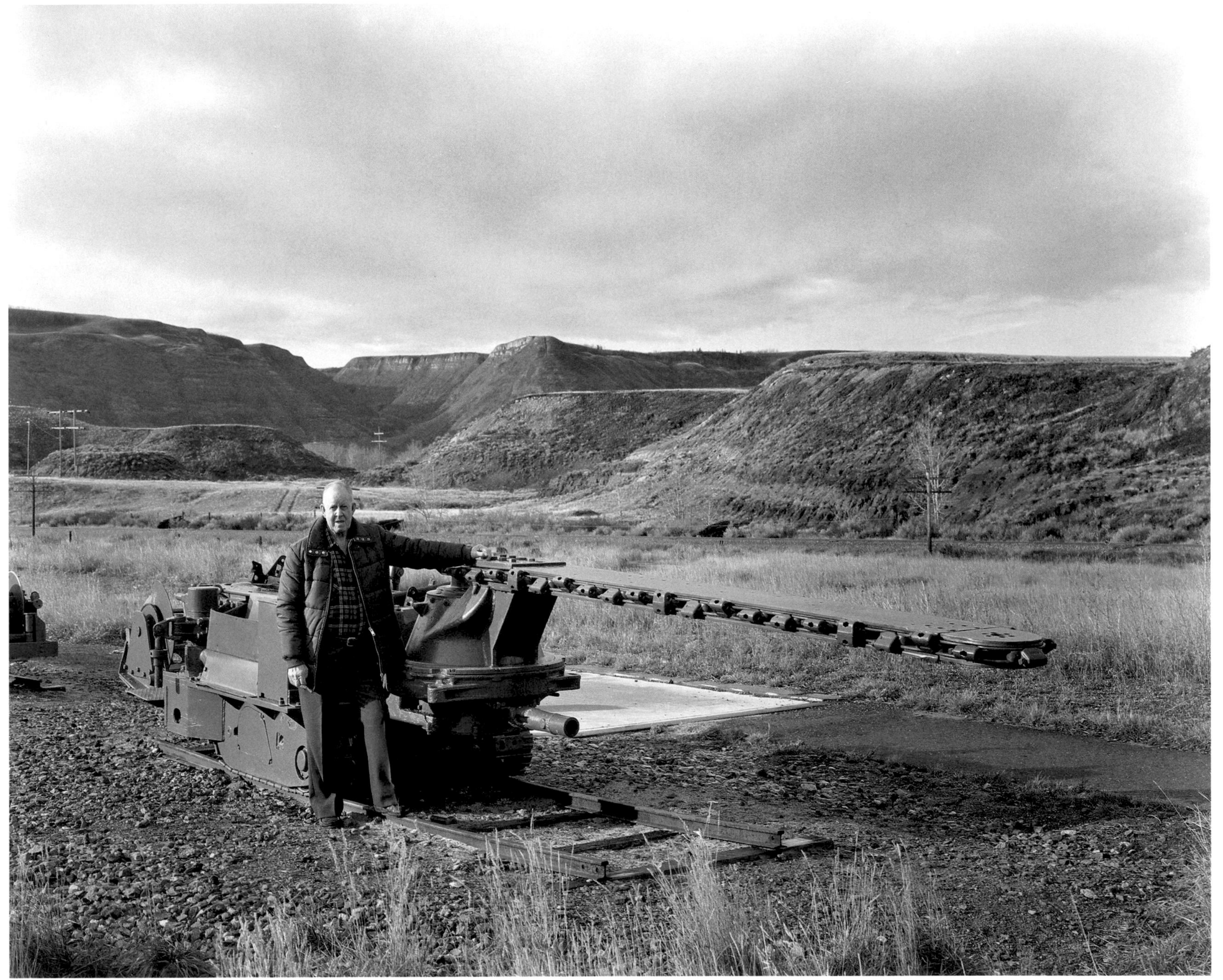

Charlie Smith, Retired Underground Electrician

In 1919, hard times on our farm forced me to work in the Evansburg Coal Mine. I was 16. I worked there until 1936, when the mine closed. Then I came down to the Monarch Mine in Nacmine. When it closed, I worked the Red Deer Mine and finished at the Atlas in East Coulee.

At one point, I took a three-year break from mining and became a weed inspector. My wife and I would take a picnic lunch and travel all around the countryside, inspecting farmers' fields for noxious weeds. We got to know all the roads pretty well.

Joe I sure enjoyed being a coal miner. You know, I'm feeling my 80s now, but I believe I could load coal yet....

I was born in 1906 in Hungary and came to Canada in 1928 to work on a farm. A friend from the old country took me underground to show me how to mine. The first time I went in the mine, by Jesus, I was white like the snow because I was afraid the roof would come down. My heart was pounding, and I was always looking at the roof. My friend said, "Don't be scared, just be quiet." He showed me how to bore and load the holes. After that, I worked in Wayne, Cambria and Rose Deer, and I finished in the Shaughnessy Mines after 27 years in the business.

Thanks to God, I only had one small accident as a coal miner. I always remembered my friend, the teacher, who said, "Safety first, always mind the roof, don't worry about the coal," and "When you lose your life, nobody is going to replace it for you."

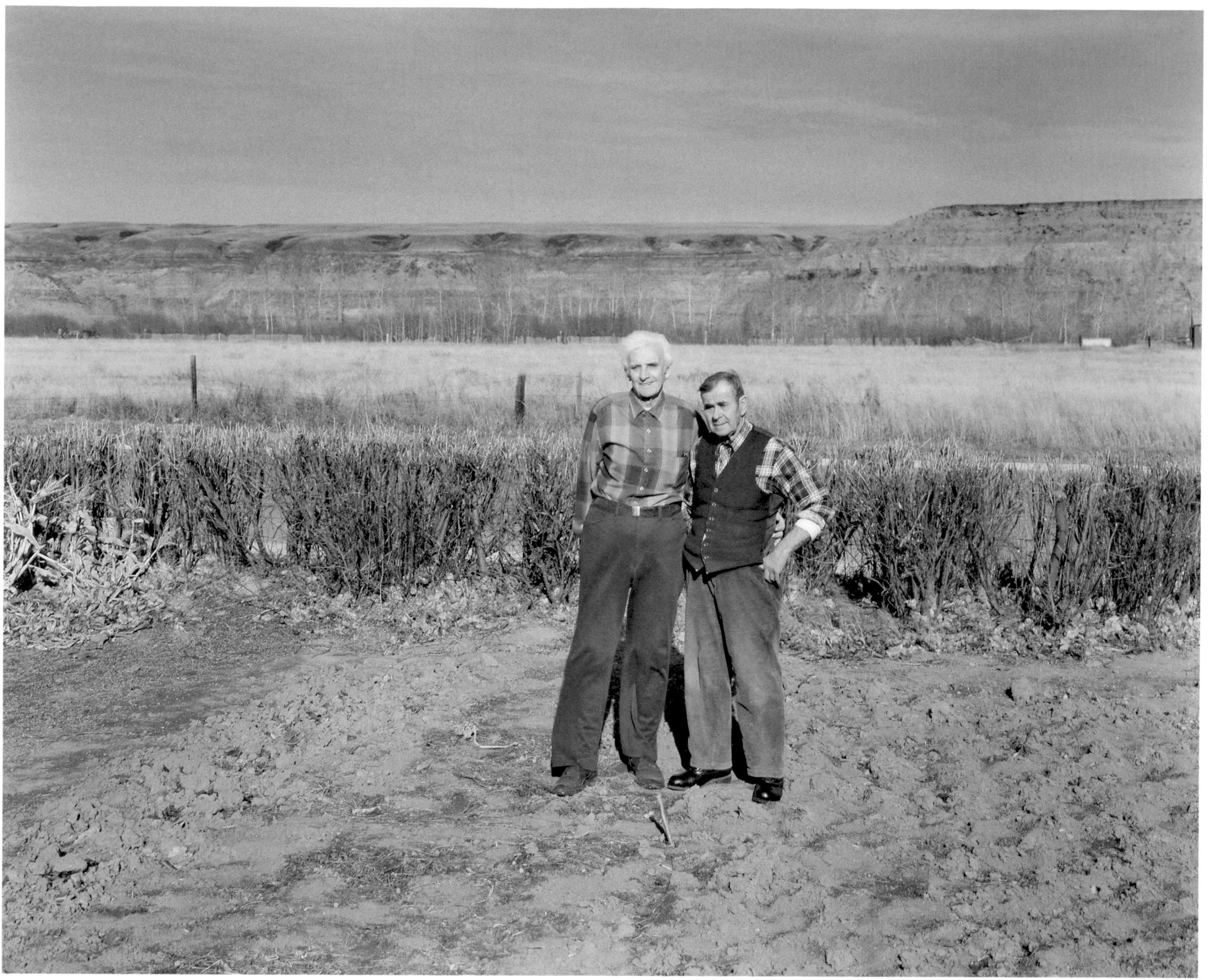

Gunner Anderson and Joe Orest, Retired Underground Coal Miners

Nick I was born in 1908, in the Ukraine. My first big job after I came to Canada was at the Livingstone Mine in Carbon, in 1929. I stayed at that mine until 1942, when I moved to Nacmine. I even moved my house from Carbon to here.

I worked at the Midland Mine for two years. In the summer, we had to walk across the railway bridge, and in the winter, we could walk across the river. Because it was a little bit closer, I got a job in the Nacmine Mine until it closed in 1961. From there, I went to East Coulee and stayed there until I retired in 1974.

I did pretty nearly everything under the sun. I started loading coal in Carbon. Then I got onto the machines. At the Nacmine Mine, I ran a duckbilled loader. I mined a lot of coal in my life to keep people warm. It was the best job in those days, but still we didn't make much money.

For 24 years, I was chairman of the union locals in Carbon, Nacmine and East Coulee. All the mines in the valley were unionized. My job involved solving any problems in the mines, including negotiating wages. My philosophy was that our members should give an honest day's work, and no one would bother them.

In the old days, every house in Nacmine had a coal miner in it. Now, there are only four men left in town who were miners. Coal mining started here about 1912. When Nacmine closed, there were more than 200 men out of work.

Nick Roby, Retired Underground Coal Miner

Rodger Bardgardmen, Kevin Fecho, Doug Grantlin, Bret Wallachuck, Tipplemen

Montgomery Coal Mine, Sheerness

Sheerness' history of underground and surface mining dates back to 1915, when an abundance of mines operated in the area and the town had a population of 125 people, a store, post office, school and a good number of miners' shacks.

Today in Sheerness, very little remains of those old glory days, although on occasion, surface mine workers will still encounter parts of the old underground mine workings.

Formerly known as the Roselyn Mine, Manalta Coal's Montgomery Mine began by supplying subbituminous thermal coal from its Sheerness site to Saskatoon's power station. The market permitted the mine to operate until the early '80s, when a minemouth power station was constructed at Sheerness. In 1984, the mine began a long term contract to supply 1.65 million tonnes of coal to the generating station, but also continues to supply domestic customers with screened and sized coal from its tipple.

All Alberta prairie surface mines must remove and stockpile the rich topsoil and subsoil before mining. At the Montgomery, powerful scrapers strip the valuable topsoil. Because these machines can move at considerable speed, individual operators require considerable skill and experience.

After removing overburden by dragline and recovering the underlying coal, the spoil piles are levelled and contoured with bulldozers. The last step in the procedure is the first step in mine land reclamation: scraper operators skillfully replace the subsoil and topsoil to promote successful revegetation.

In the prairies, mined land can usually be returned to agricultural use within three years after mining has been completed. It is not unusual to hear miners claim that by improving land drainage on swamp or muskeg properties, they leave the mined land in better shape than they found it.

Ken Stotz, Scraper Operator

Art Ralston, Retired Surface Coal Miner

Art We came to Canada from Scotland in 1905, the year I was born. My dad had been a store keeper, so he bought an old log building here and started a store. When the railway came through in 1908, he moved the store to a place called Lignite — but that town never did amount to anything.

I grew up beside the mines, and my first chance at a paying job was in the Sissons Mine. I started working for Eric Sisson's grandfather in 1922 when he first used horses for stripping the overburden. It took all summer to get just a little bit of coal. We used wheelbarrows and shovels to clean and load the coal. It was all hand work, including the drilling of holes with a breast auger to blast the coal out. During the Depression, I worked at the mine for a dollar a full day, and I had to board myself. Young people think I'm dreaming when I say that.

I said I wasn't going to work in a coal mine all my life, but that's what I ended up doing. I kept working until I was 80 because I had an easy job on the scales and I felt like working. I could also run the tipple or drive a truck.

Then, when I got to be 80, I finally said to my employer, "Eric, doggone it, don't you think I should quit?" He said, "I don't know why you should." "Well, I'll tell you, Eric, I think I'm going to retire because, doggone it, Eric, it's interfering with my fishing."

Fishing was always my hobby. I have a friend in town, and we've been fishing together 60 years.

Eric My grandparents came from England to Naniamo, British Columbia. My grandfather got in trouble with the unions there — told them all to go to hell — so the family moved to Alberta.

During the early 1900s, there were quite a few underground mines in this area, particularly along the banks of the Red Deer River. A mine shaft was sunk in my grandparent's farmyard, next to the CN Railway.

The Sissons Mine has been in operation since 1922 — we just celebrated our 70th anniversary.

When my grandfather passed away, my dad took over everything. Later, I worked the mine during the winter and farmed our three sections during the summer. And at that time, we had a mine manager working and living here. We were reclaiming our farmland long before the government told us to do it.

Both my grandfather and father did surface mining here. I do the same, but I use a contractor to take off the overburden. The seam is only three feet thick, but the heat content is high at 10,000 BTU per pound.

Later, my brother took over the farm and I took over the mine. I've been in this mine for 32 years, and I'm tired even though I'm only 50 years old. The problem with running a small mine like this is that I have to do everything: I do all the welding, I rebuild the motors myself. If there's something I can fix rather than buy it new, I do it. You get tired of this when things break down and it's 40 below.

Eric Sissons, Operator

Roy I look back with a great deal of pride on my days as a mine manager. Once, when we were loading coal, I had to go to a meeting in Calgary. We were behind on orders, and pushing everything to the limit. But when I came back, the men had loaded more coal in a day than ever before. I discovered later that they'd worked overtime to get an extra car loaded just so they could kid me that the mine ran better without me. To me, it was extremely satisfying to have a crew that would do that.

I was born in 1918 in Saskatchewan, where my father was a homesteader. We moved to Alberta when I was two. When I was 29, I decided to take a job at the Vesta Mine for a year — it ended up being 29 years.

I became interested in mining by doing electrical work, and got my mine electrician's papers largely at the encouragement of the Mannix Company. Then, as time went by, I got my mine manager's papers, and managed the mine here for 13 years. Later, I became the training co-ordinator, and did that the last two years of my time with Mannix.

The coal industry was always challenging, particularly when we switched from the old friction machinery to the electrical machines. And I liked the people part of the coal industry. It was a matter of personal satisfaction to work with my crew, sometimes under difficult conditions. You worked with them, used their ideas, and let them exercise initiative.

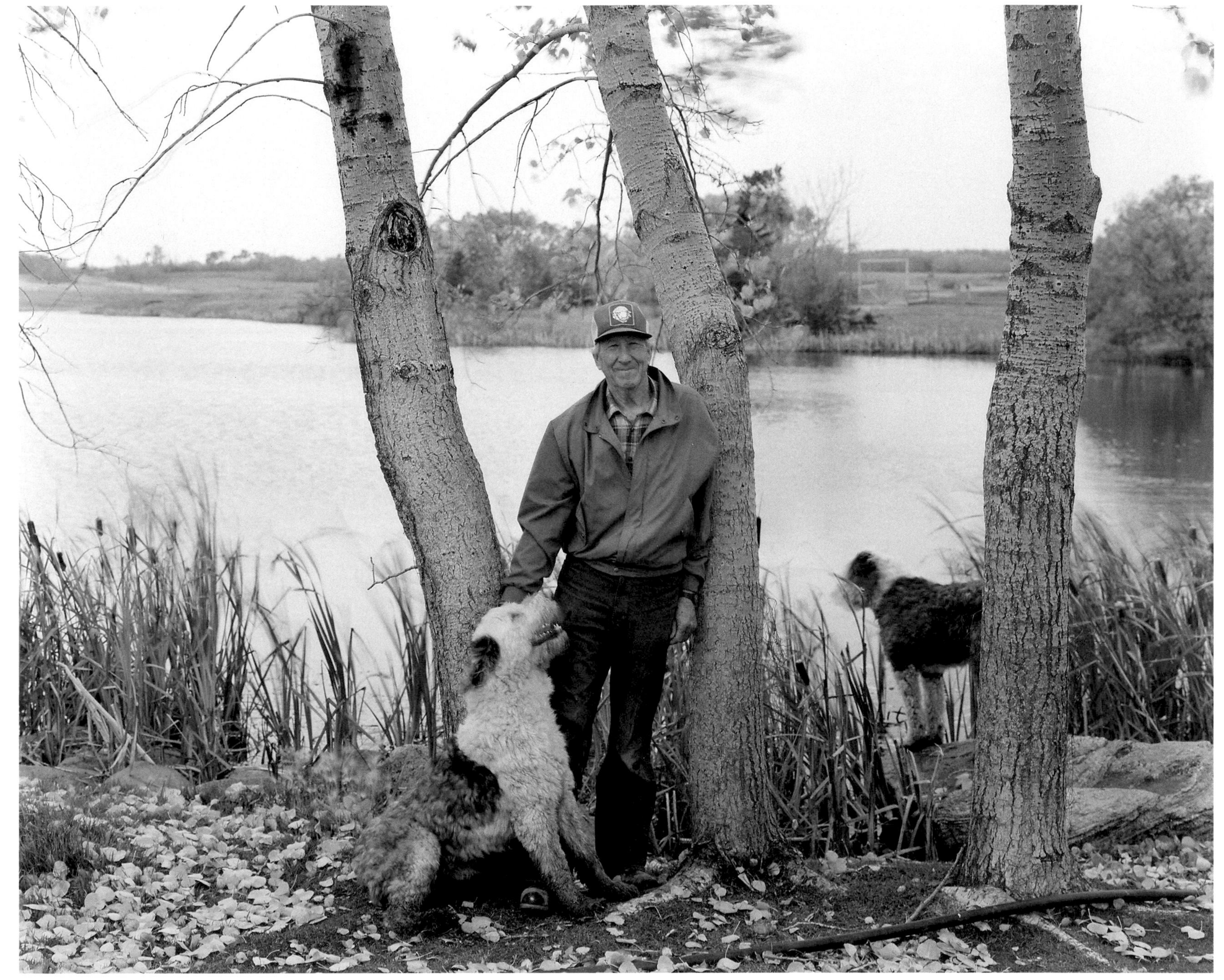

Roy McBride, Retired Surface Coal Mine Manager

Louie Thibault, George Rowland and Blake Taylor, Retired Surface Coal Miners

Louie I was born in 1919, in Halkirk. During my later years, I rented the farm out to the neighbour and worked at the mine steady until my legs gave me heck and I had to take early retirement.

I started at the mine full time, eight hours a day, five days a week, and it suited me very nicely. It was more fun when the tipple was going. An odd-ball customer would come in and you'd give him hell. We used to get some awful mixups around that tipple — especially me. Oh, that was the coldest place in the world — oh my.

George For 12 years, I loaded many thousands of tons of stoker coal on boxcars. I really enjoyed it during the last few years when I'd retired from the farm. When I had both jobs, the farm and the mine, it was too much.

Well, hockey is probably one reason why I wouldn't go south in the winter. I like to be involved with kids and help them as much as I can. Maybe I ain't got many years left, who knows!

Blake I was born in 1918 in a little sod shack about a mile east of Halkirk. My dad, who was a pioneer, cut out his own homestead.

In 1953, I started at the Vesta Mine and have been here ever since. I loved every minute until I hurt my hand. After that, I was a damn security guard, but it was a struggle. I was never a boss, but still, I was happy.

During the summers, I farmed, and I went broke farming, too. The job with Manalta was always the one I took care of.

During the early 1900s, underground mining in the Halkirk area was extensive. The room and pillar mining method was used in the region until the 1940s, when surface mining became widespread.

In 1948, Manalta Coal predecessor Alberta Coal bought the land that now encompasses the Vesta Mine. In '56, Canadian Utilities constructed the first unit of the Battle River coal-fired power plant. As this plant increased in size, so did the Vesta Mine, producing 1.7 million tonnes of subbituminous coal annually.

Vesta's mining sequence begins with overburden removal using a large dragline. Production miners in the pit use shovels and loaders to recover the coal when the seam is exposed, and truck drivers haul the run-of-mine coal directly to the Battle River Plant.

Long before provincial laws stipulated that mined land be recovered with reclamation techniques, the Vesta was conducting its own program of reclamation. The site has been used for a joint industry-government research project to examine reclamation methods.

The mine's dragline operator and oiler rightly deserve their prominence in prairie surface mining operations, as their performance is often critical to the mine's overall operation. If a dragline breaks down or does not perform to its capability, the rest of the operations either slow or come completely to a halt.

The job of dragline operator demands a rigorous apprenticeship. Most operators begin by first learning to use heavy equipment like bulldozers, then moving on to become dragline oilers. Oilers start training on the dragline with 15 minutes of practice operation per shift. After a year, the oiler may relieve the senior operator for up to an hour at a time.

Good dragline operators usually have a natural talent for the timing, co-ordination and "feel" of mechanical equipment. This position is one for which there are few vacancies.

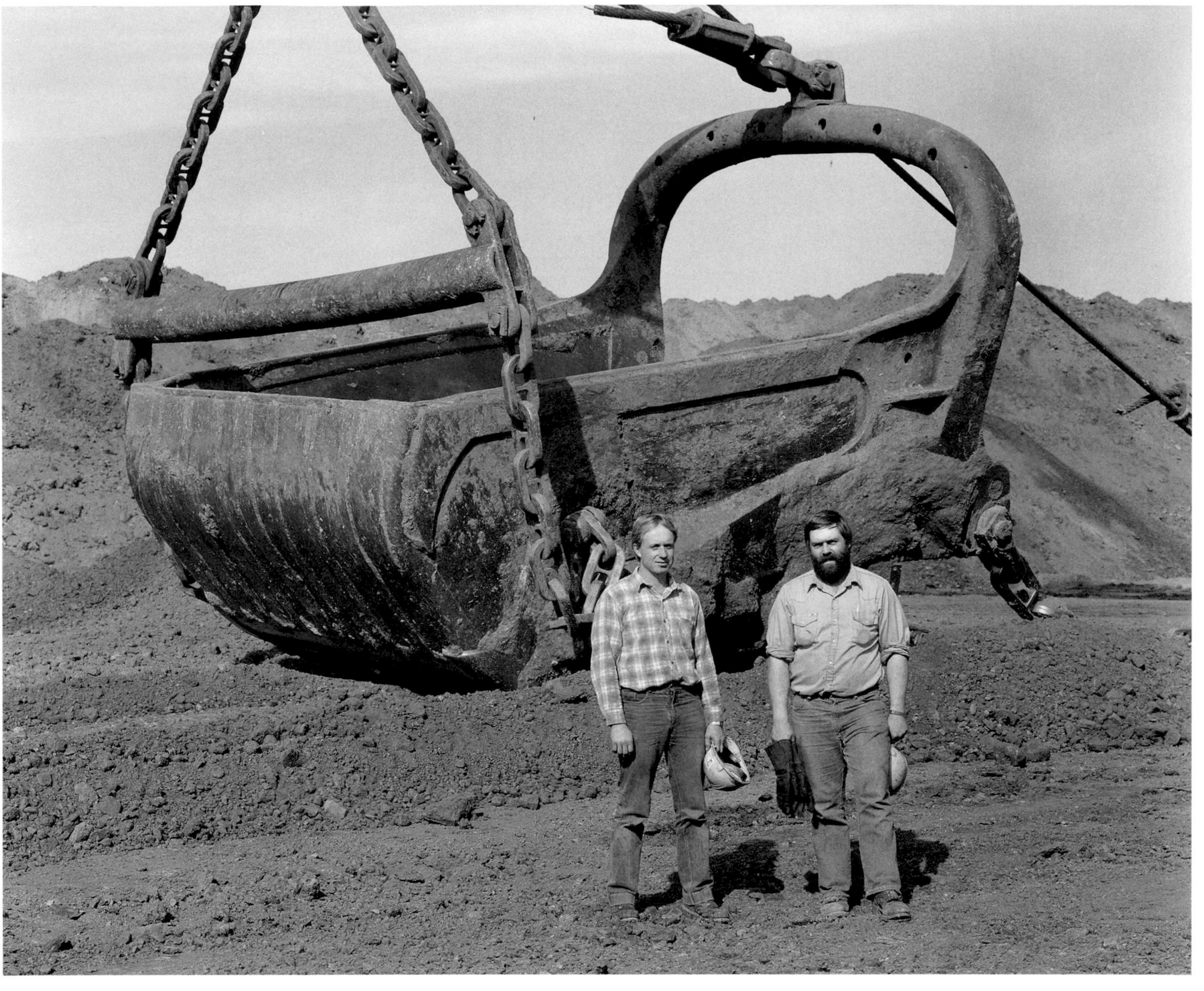

Bruce Jackson, Dragline Operator and Tim Neilson, Dragline Oiler

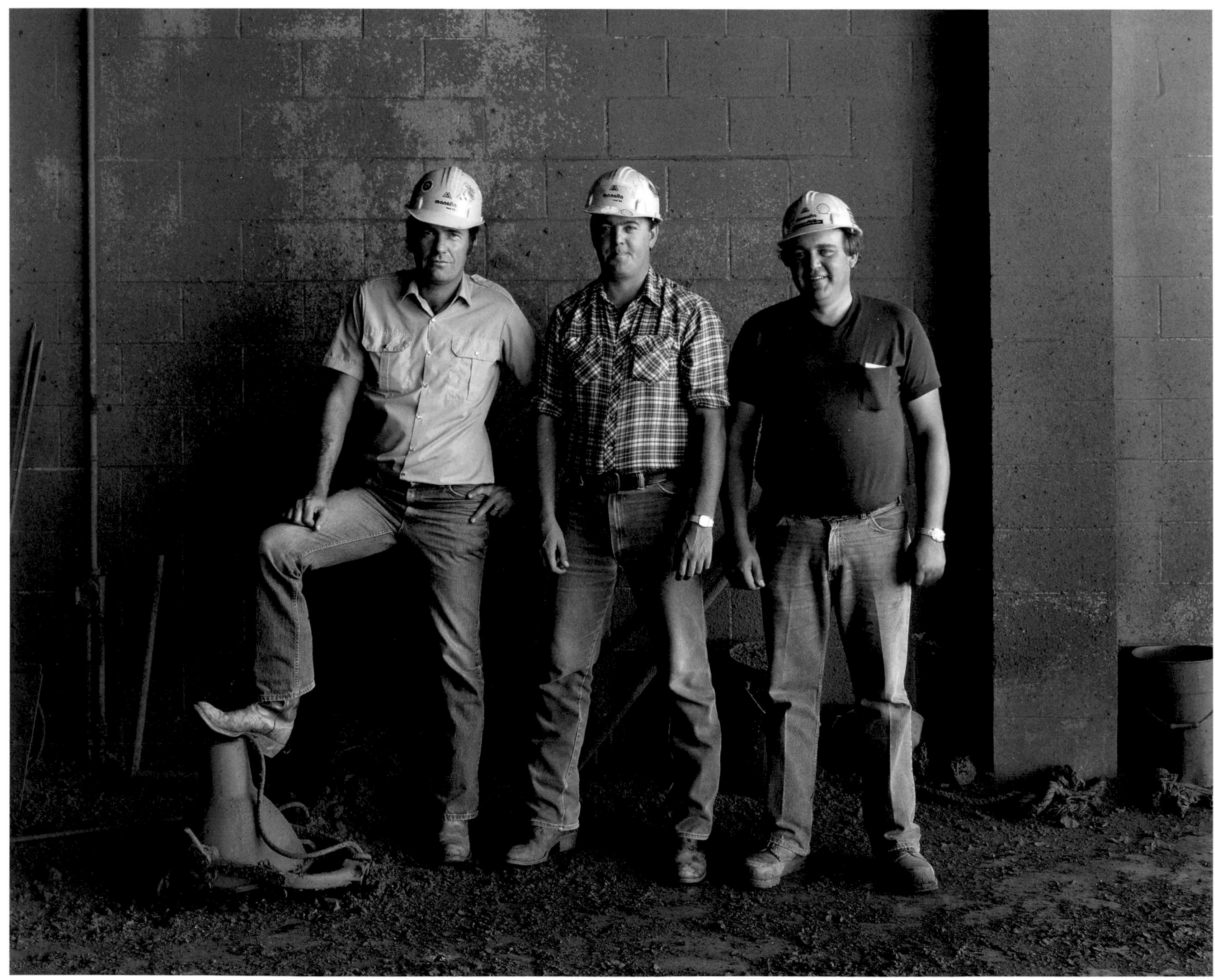

Wayne Venance, Pit Supervisor, Larry Schroeder, Pit Foreman and Pete Popowich, Pit Foreman

Paintearth Coal Mine, Forestburg

Ron Stenson, Pit Foreman

Paintearth Coal Mine, Forestburg

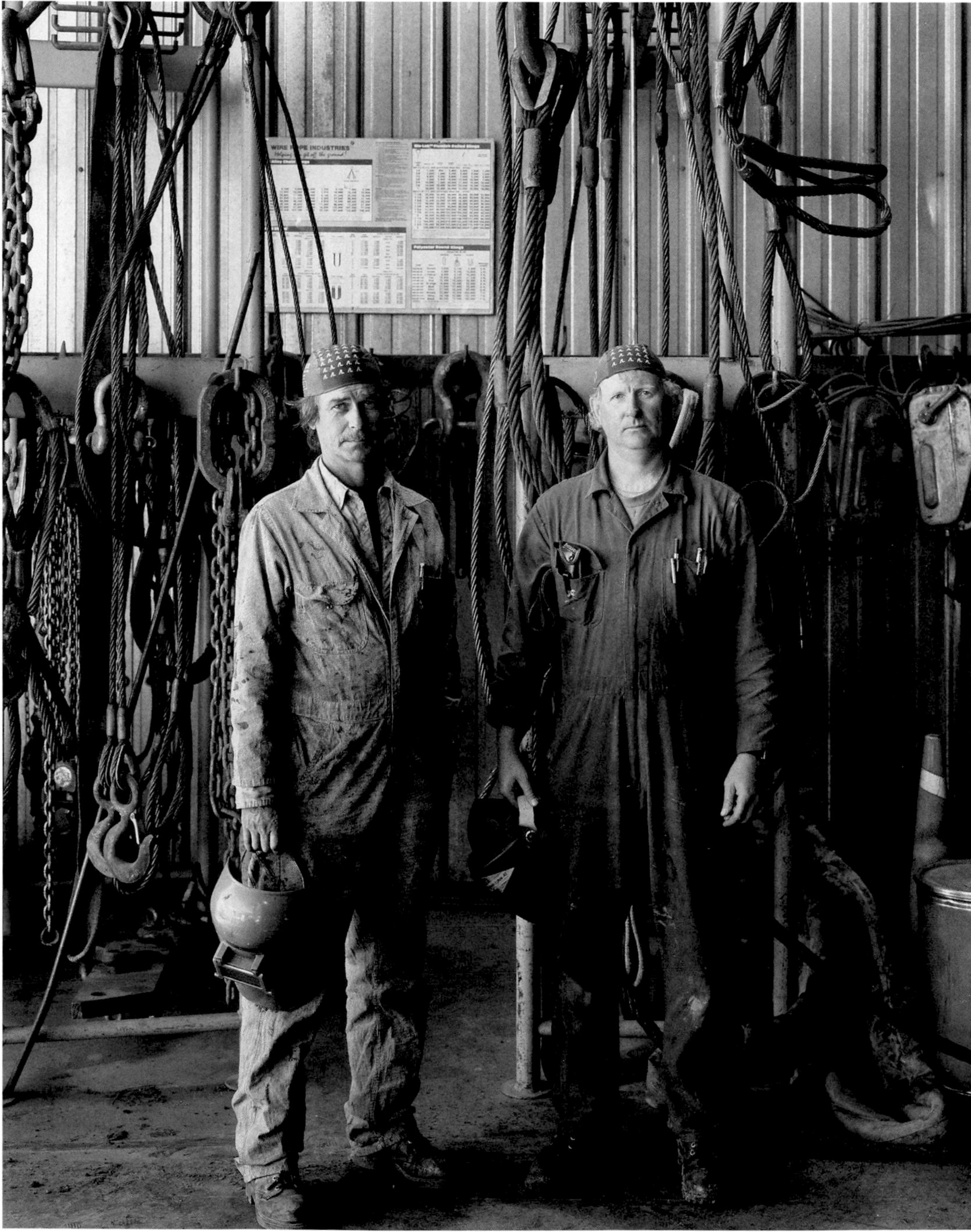

Ron Fossen and Jerry McMahon, Welders

The Paintearth Mine had its origins with the renowned Bish family. In 1907, the Bish brothers established a small underground coal mine on the slopes of the Battle River Valley, not far from the existing Paintearth Mine. It was the largest and most sophisticated facility in an area flourishing with mines. At peak points, up to 25 mines were operated in the area by local farmers, who either kept the coal for personal use or sold it to neighbours.

Financial difficulties forced the sale of the Bish Mine, and Luscar Coal and a group of other mining interests took ownership of the site. In 1950, this group formed Forestburg Collieries and opened the Diplomat Mine which was the first large-scale surface mine in the region.

In 1956, Luscar obtained full ownership of the Diplomat and arranged a long-term contract to supply coal to the Battle River Generating Station. When coal reserves at the Diplomat neared depletion in the late '70s, Luscar replaced the Diplomat with the Paintearth Mine. The Diplomat operated for over 30 years, and the Paintearth is expected to have the same mine life.

Situated across the river from the old Diplomat, the Paintearth is a captive mine with total production committed to the Battle River Generating Station. Nicknamed "Brutus," the mine's dragline strips the overburden after removal of topsoils by scrapers. Flat-lying subbituminous coal seams are then recovered using loaders and trucks that haul the coal directly to the power station.

As his name implies, the pit foreman oversees the pit operations. It is his responsibility to see that the coal moves continuously to the Battle River station at consistently high levels of safety and efficiency.

Keith I was born in 1943, in Sheffield, a coal mining region in England. At an early age, I was exposed to surface coal mining within walking distance of our house — in fact, the strip mine was on our street.

I got familiar with the different machines and with the operators who would give me rides on their vehicles.

Later, I got a degree in civil engineering from Sheffield University. While I was a student, I worked summers at the mine on my street.

In 1974, I came to Canada to work on Luscar's Coal Valley Mine. My first job was to supervise the mining of a large bulk sample for testing by Ontario Hydro.

In 1981, I became manager of engineering at the mine; then in 1984, project manager of the proposed new Sheerness Mine. That mine didn't get off the ground, so I became project manager at the Paintearth Mine — and I've been in Forestburg since then.

Now, as chief engineer of Forestburg Collieries, I am involved in planning work — short and long-range plans including everything from overburden removal to reclamation.

I've never lost my fascination for heavy equipment. My hobby is photographing and documenting mining equipment, particularly old mining equipment.

And there's more old equipment here in Canada than in England. Back there, the scrap business is very strong — you don't see old machines lying around. But here, you can still see the old '40s surface mining equipment in Forestburg and Sheerness.

I've written many magazine articles on the history of mining equipment. One of my proudest accomplishments was to co-found the Historical Construction Equipment Association, based in Ohio. Since the formation of that group in 1985, I've been the editor of its magazine, *Equipment Echoes*.

My pride and joy is a fully restored 1941 10B Bucyrus-Erie dragline, bought in Edmonton. This is one of the few mining equipment antiques of this type in Canada.

Keith Haddock, Mining Engineer

Ken Norman, Reclamation Operator

The Rosalind quarry was originally part of the Magna Cove Barium Corporation, and was eventually purchased by Dresser Industries. The large oil drilling company of Haliburton also took an interest in the quarry, and the resulting joint venture renamed the project MI Drilling.

Bentonite clay was first quarried at Rosalind in the 1950s. In the following decades, the quarry provided a significant amount of drilling mud for the oil industry, but this market dropped off when American clays entered the competitive marketplace in the '80s.

The plant subsequently switched to production of industrial clays for fertilizer, refractory, agricultural and construction industries, particularly for producing refractory bricks in eastern Canada.

During peak production, the operation employed up to ten people to dig the bentonite, dry it and package it for shipment.

The quarry officially closed at the end of 1992.

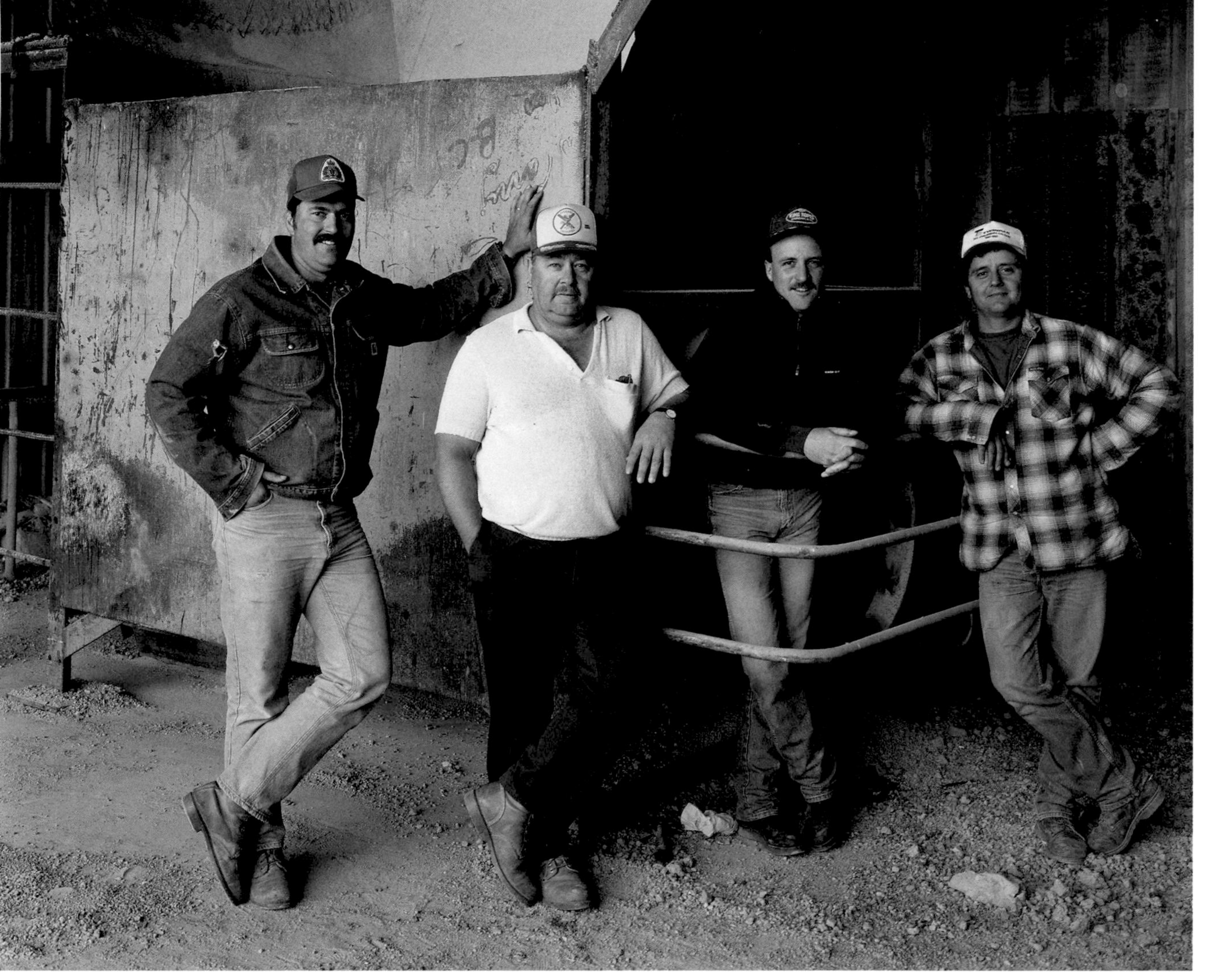

Jack Helmig, Alvin Vos, Gord Jessiman and Doug Volk, Quarrymen

Dodds Coal Mine, Ryley

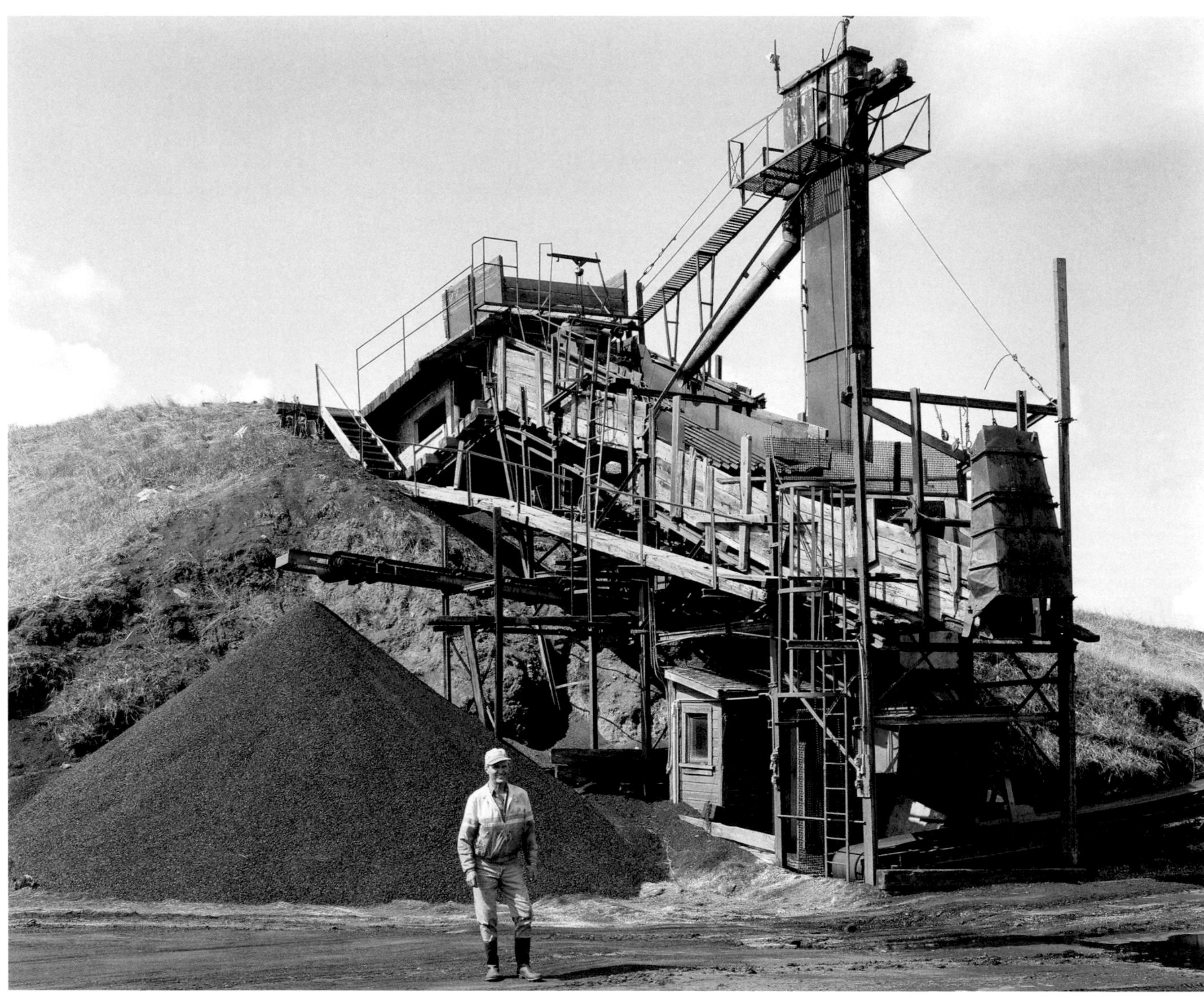

James Nordby, Operator

James The Dodds Coal Mine is one of the oldest mines in Alberta. It started in 1905 as an underground operation, and there were 62 men working here. We leased Dodds in 1952 because my friend came to me and said, "Let's give coal mining a try." In 1962, we bought the mine.

When we first started, the Black Nugget Mine was operating nearby, and there was another mine close to here where, in the '20s, they used to strip with mules.

Our coal seam varies in thickness from 5.5 to seven feet, and is between 16 and 26 feet deep. We sold mainly nut and pea coal, and it's good coal, with an ash between six and eight per cent. Our sales averaged between 15,000 and 25,000 tonnes a year. Black Nugget, which was the larger mine, sold anywhere between 40,000 and 60,000 tonnes a year. When we first came here, coal was worth only $2.50 a tonne; now, it's $22. The trade name for this coal is Red Flame.

I've leased the mine out to my two sons-in-law. But I pester them and I still give them a little advice on the stripping. If you start moving dirt twice, it can cost money.

I operated this mine for 35 years, and I wouldn't have stayed here if I hadn't enjoyed it. I was my own boss, doing the stripping and running cats. In fact, I did better at mining than I did at farming. It gets in your blood, you know.

Corey, Dana, Brad and Peter Kudrowich, Family Operators

Dodds Coal Mine, Ryley

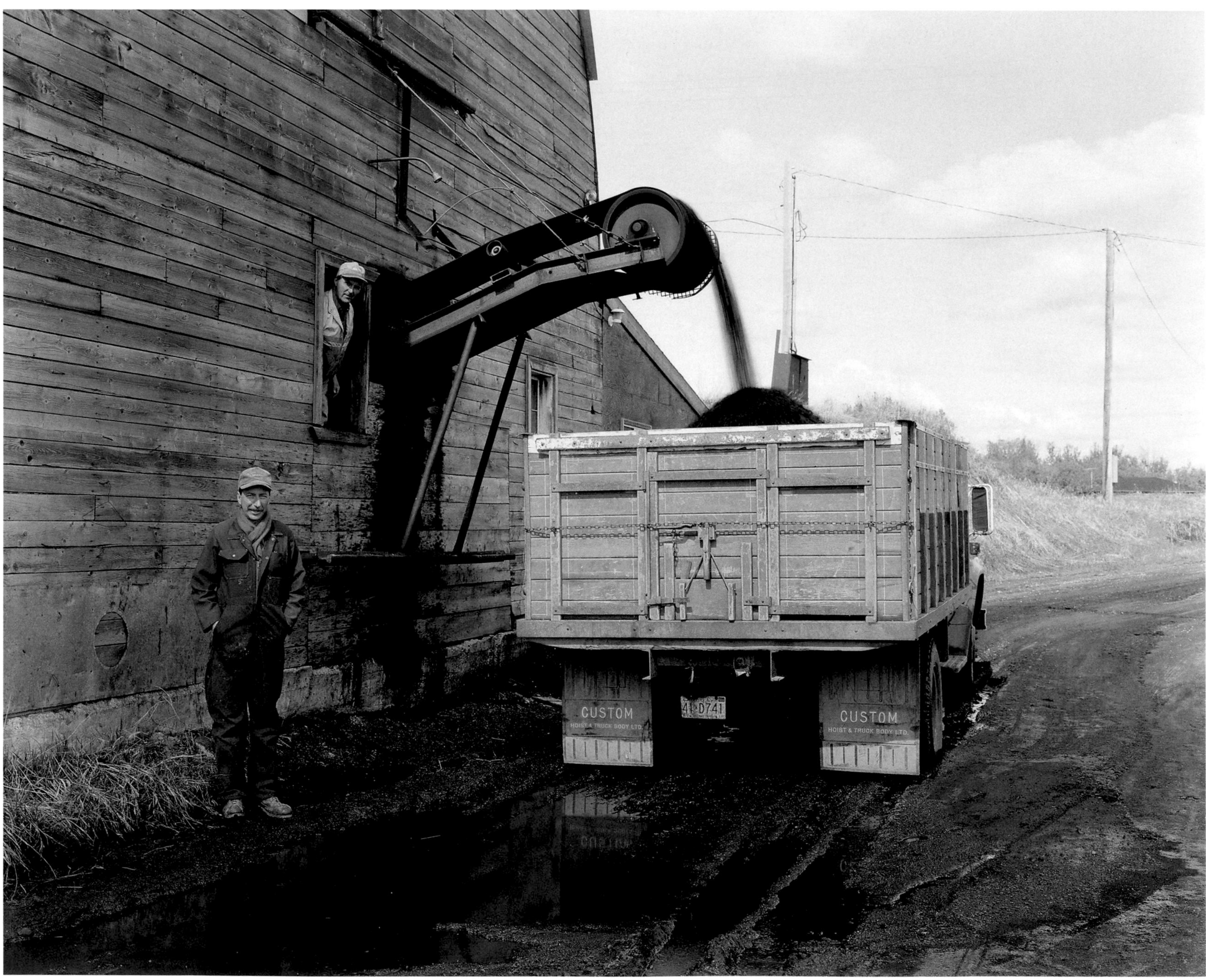

Lorne Fill, Farmer and James Nordby, Operator

Ray Originally, this quarry was mined to make the ballast for the railway when they put the coal mine in. It started out as a hand operation. The railway was just at the edge of the quarry, and the mine was only a mile away. The old timers tell me that down at the bottom of the hill there are the remains of a turntable where they used horses to crush the limestone.

I got started in this limestone quarry through Fish Creek Construction. We sublease it from Westmin. The limestone is sold directly to Limeco in Rocky Mountain House. For lime production, it's very good quality — it's used for cattle feed and for neutralizing sulphur. Lately, we've crushed up some limestone for the sugar beet industry in Taber. We've also sold large quantities of this limestone for river ballast because it's acceptable environmentally.

I do everything from sales to driving the loader to driving the trucks. We get guys out as we need them to do the drilling, blasting, and crushing.

Since the coal mine closed down, the old timers have established a camp in Nordegg — like a trailer park. They all come up here during the summer time, sit around and tell lies. They have even rebuilt the old golf course that existed when the Nordegg Coal Mine was in operation.

Ray Borley, Quarryman

I was born in Nakusp, British Columbia, in 1914. I started out loading boxcars at Nordegg. The company found out I was shoeing horses in the evenings, and they put me in the blacksmith's shop. That's where I stayed until the mine shut down in 1952, after a fire burned everything except the tipple. After rebuilding, Nordegg operated for a couple of years, and then went broke.

Nordegg had about 65 horses they used underground, and another 10 they used in a bush camp to tow out mine timbers. The mine didn't have underground stables — the horses were brought out each shift. Many teamsters didn't know horses, and didn't give a damn. I saw horses come out with welts on their backs. The miners pounded the buggers. You see, not every horse wants to work in the mine. Not every man wants to, either.

I sharpened the pick and rock drills for the miners. Most miners owned their own hand picks, but the company bought the air picks and jackhammers. Some guys would come in and give us hell for not tempering their picks correctly. If you tempered them too hard, then they'd bust.

After Nordegg, I spent most of my time in logging camps, and blacksmithing. I was a lumberjack, but I never worked around the mill — that's one place I wouldn't work for anybody. I'm all nerves around a sawmill.

As far as I'm concerned, Nordegg was the nicest place I've ever worked. I never did want to leave there, but had to because the elevation was too much for me.

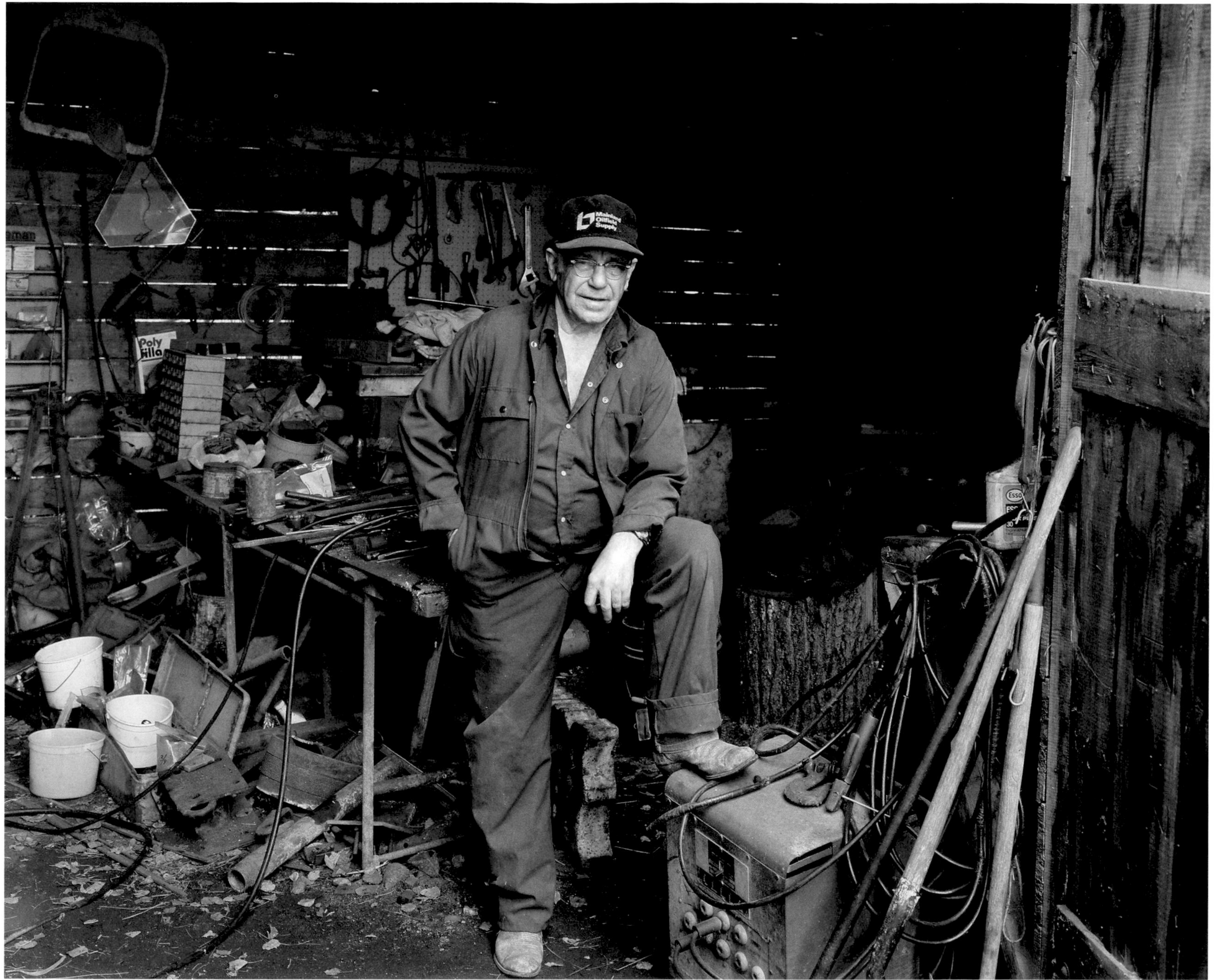

Louis Niedermouser, Retired Blacksmith

Coal Valley Mine, Coal Valley

Coal Valley is one of several mines situated in the historic Coal Branch area of Alberta. Originally, the term "Coal Branch" was applied to a railroad spurline built by the Grand Trunk Pacific, but later the name was taken over by the Canadian National Railroad.

During peak mining times in the region, Coal Branch towns like Mountain Park, Cadomin and Mercoal each accommodated more than 1,000 people. Many company-owned mining camps in the area were large enough to officially be recognized as towns: Robb, Coalspur, Foothills, Coal Valley, Lovett, Mountain Park, Cadomin, Mercoal and Luscar were large, bustling hives of the coal industry. Today, most coal miners from the Coal Valley Mine in the area live in Edson.

During the '30s and '40s, the Coal Branch boasted 14 coal mines, with the product's primary market being the steam locomotive industry. Production here reached its peak in the 1940s when more than 1.5 million tonnes of coal were produced annually.

Most of these were underground mines, but in 1918 some companies began surface mining the deposits, using steam-operated draglines and shovels. The Coal Branch was one of the first coal mining regions in Canada to use this methodology.

By the mid-50s, almost all mining operations had ceased in the Coal Branch. The Mountain Park closed in 1950, Cadomin in 1952, Coal Valley in 1954, Sterco and Robb in 1955, and finally Luscar in 1956.

In 1970, Luscar Ltd. opened a mine to supply coal to Ontario Hydro, and the first unit train of high volatile bituminous coal left the mine for Ontario in 1978. This shipment would prove an important milestone in developing western coal in eastern Canada markets.

Today, new offshore markets have encouraged the mine to expand its production to two million tonnes annually. The mine's old-timers would no doubt be shocked to see the massive draglines, trucks and shovels digging through the old underground workings that they had once dug — by hand.

Dale Dunlop, Dragline Operator

Back row: Wally McMorran, Materials Manager, Bob Latimer, Engineering Manager, Morris Ennis, Plant Manager, Ron Stard, Mine Manager, Howard Ratti, General Manager *Front row:* Jim Smitten, Controller, Al Hammernick, Human Relations Manager

Don Ofner, Driller

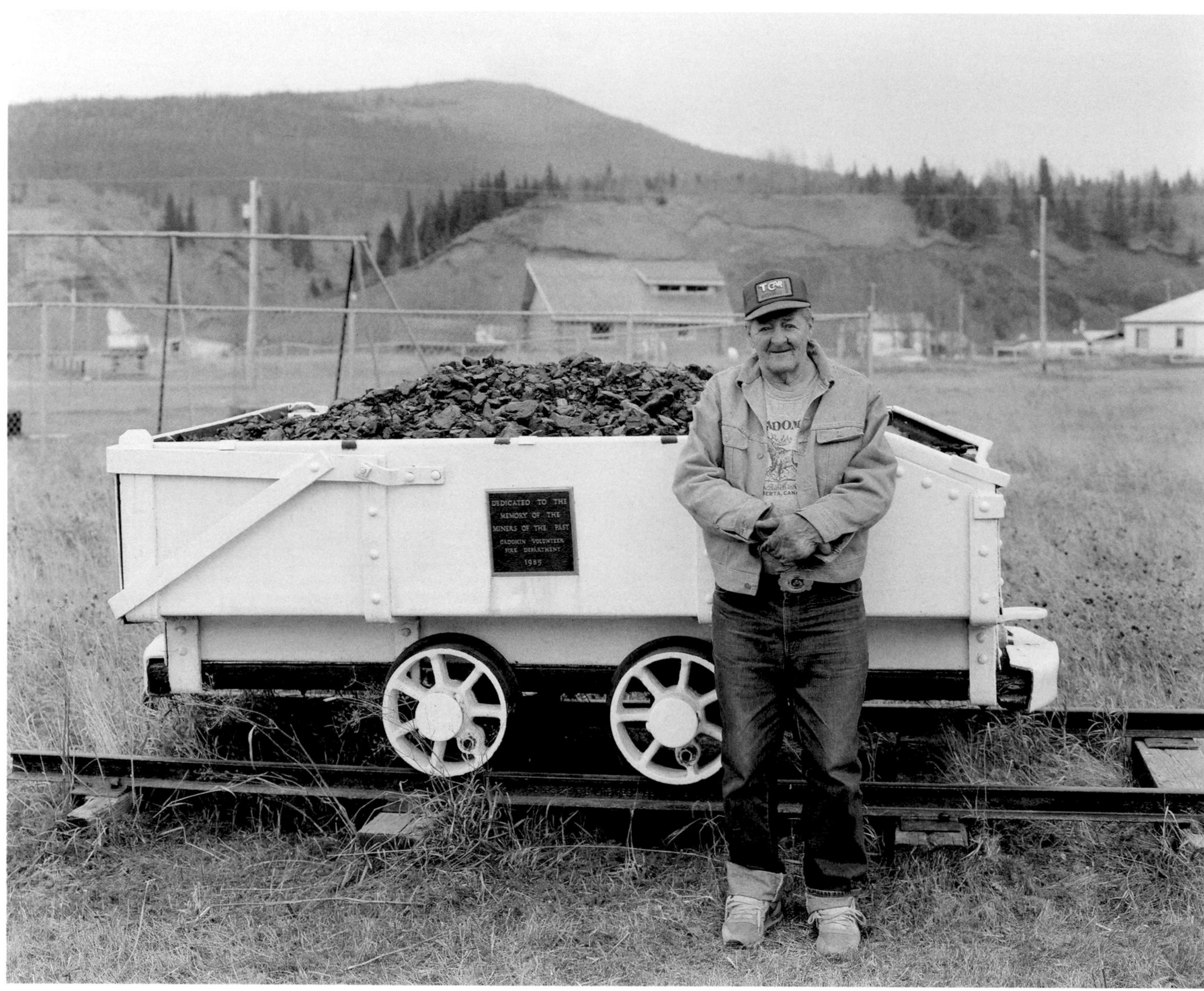

Jack Kazmir, Retired Underground Coal Miner

Jack I was born in Manitoba in 1912. When I was drafted into the army during the war, a friend told me I could get out if I was a miner. And although I wasn't a miner yet, I told them I had done similar work, and could get a job in coal mining if they let me out of the service.

I ended in Coal Valley where I ran a cat and scraper for Mannix. The company had a contract to strip coal for Cadomin Coal. Mainly, I worked on a 2.5-yard Buycarus Erie Shovel. Boy, the dust would blow, it was cold and the old levers got pretty heavy by four o'clock. After the Cadomin mine closed, I worked at the Luscar Mine until it too closed. Then I joined up with Inland Cement in the Cadomin Limestone Quarry — I worked there for 21 years.

For a time, I worked on top of the mountain, but I couldn't really stand heights. If there was a $1,000 bill on the edge of the cliff, I wouldn't have been able to pick it up — it scared me to death. So most of the time, I loaded limestone into rail cars for shipment to the Edmonton cement plant.

When we first lived in Cadomin, I remember the wind blew so hard that the outhouses were blowing down the street and the walls of our house were moving back and forth. We were so scared that my wife and three kids hid in our cellar below the house.

William I was born in 1912, near Bellis, Alberta. During the Depression years, it was impossible to get a job. I'd won prizes in school for my drawings. As an artist, I started painting mountain scenery on pieces of wood and selling them from door to door. Somebody said to me, "You go up to the Coal Branch and you'll sell some there." So I came up here on a freight train the night King George V died, in 1936. I couldn't sell any of my pictures, so I got a job wheeling ashes in the Cadomin Mine. Then after that, I went underground, bucking coal.

When I was working in the mine, a guy who took a liking to me talked me into taking up photography. I started up with a box camera, and before I had a chance, everybody was after me to make pictures. In those days, people wanted postcard pictures so they could put them in a letter. I did my own processing. In the army, I was stationed in the Photographic Branch in England, where I learned all about printing and making half-tones. When I came back to Cadomin after the war, I decided to start up a studio here with the money the government provided to get re-established.

But when the mine shut down here, everybody left Cadomin. They left their houses, they even left their horses. That's when I started making movies. There was an article in Popular Photography Magazine about movie-making, and that's how I learned. My first movie was a 3-D film called *Whiskers*, then I went into Cinemascope. One of my films, *Dawson City Joe*, got into the National Archives, and was shown at some theatres in Alberta.

I always thought that if you made a movie, the money was going to roll in. But it didn't work out that way. I always had to rely on outside money from the mine.

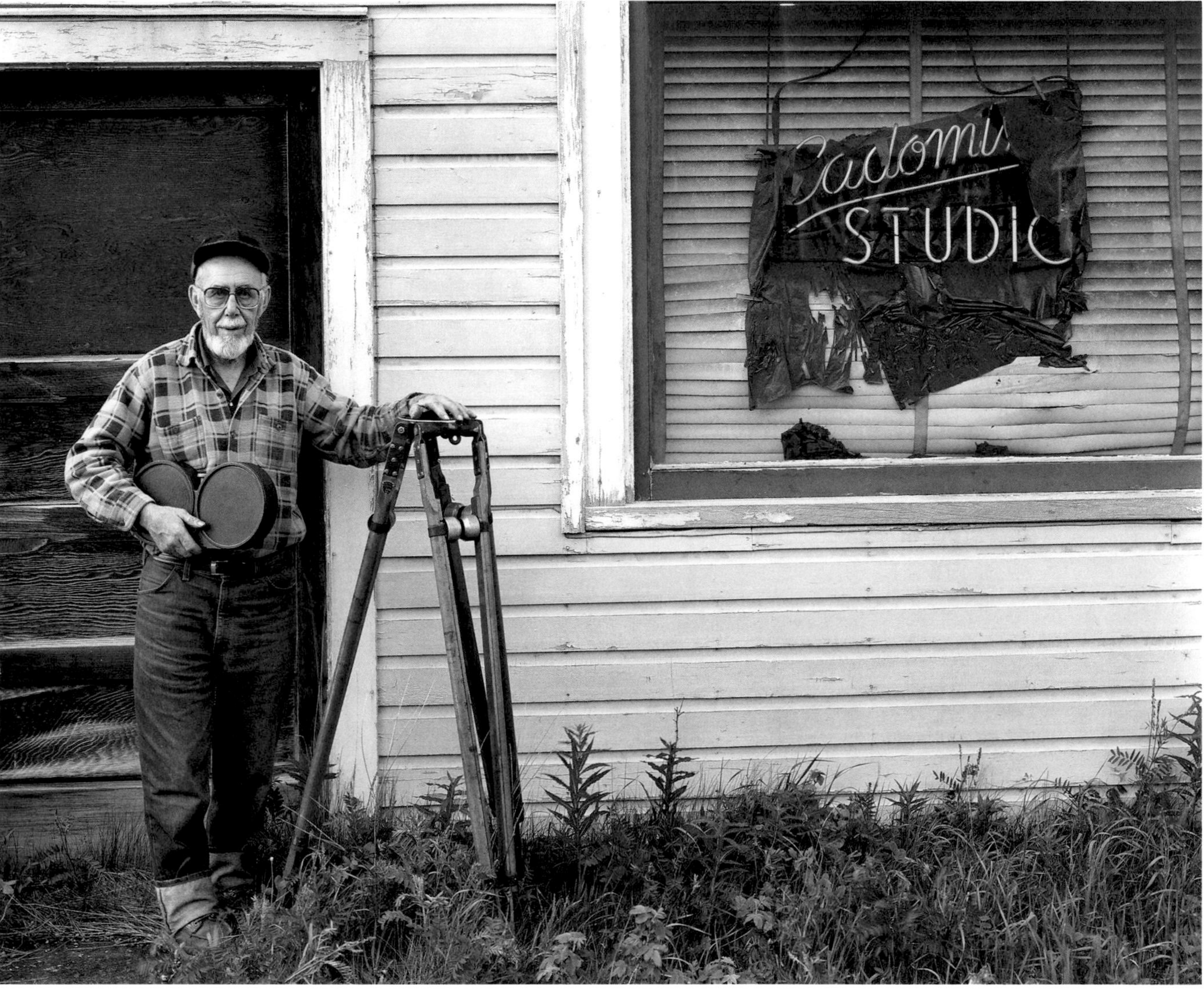

William (Flash) Shewchuck, Retired Underground Coal Miner, Photographer and Film Maker

Bud I never worked underground, because I figured life was short enough without working down in the dark with that headlamp. But my dad and grandfather would rather have worked underground than any other place.

My grandfather worked in the coal mines of Nova Scotia when he was 11 years old, and he finished at Mountain Park when he was 75. He lived until he was 98, and he couldn't read or write. My dad's folks came from Springhill, Nova Scotia, to the coal town of Brule, in 1914. I was born there in 1927.

For 20 years, I was at the Cadomin Quarry. First, I was head of the union, and then I became a supervisor. I did everything in that quarry — I ran the cat, the shovel, the drill. When I was foreman, I often worked long hours to produce rock for the next train. Guys I knew were scared to death of the height of the quarry. With me, it didn't matter. The front of that mountain was over a thousand vertical feet — a straight wall. I'd take a cat and push the back rock over, and then hang the blade of the cat over the edge all day long, back and forth.

Ten years ago, I had open heart surgery. Many people said to me that they'd think twice about having open heart surgery. I said, "What is there to think about? Either get it done or die." And I haven't looked back. In fact, now I have a little girl. I'm 64 and she'll be seven in a few weeks.

I haven't but a grade nine education. I've been to the school of hard knocks.

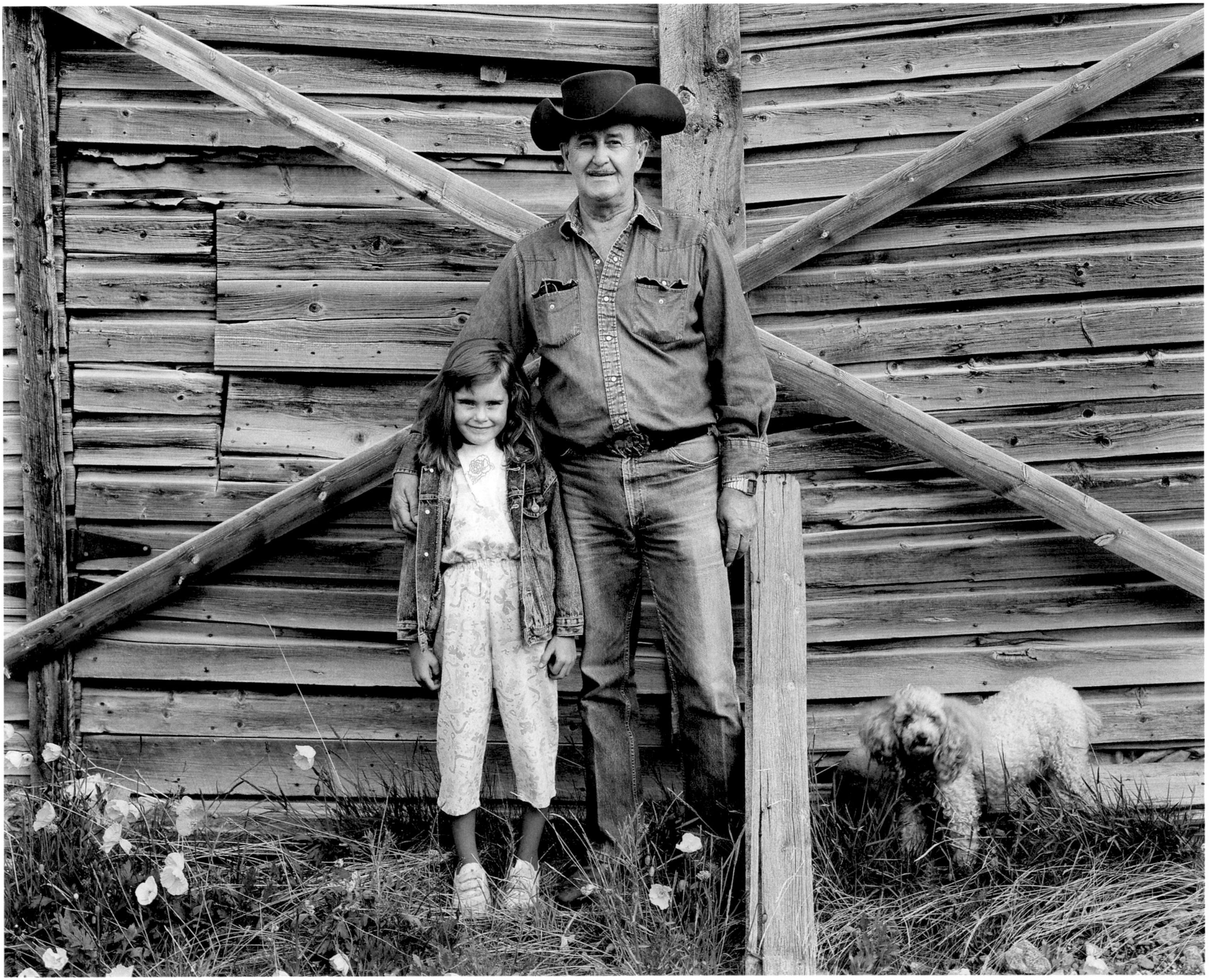

Retired Limestone Quarryman, Bud Malloy and daughter

Cadomin Limestone Quarry, Cadomin

A small family-run company first established these limestone quarries, producing lime in small furnaces. After a few ownership changes, Inland Cement purchased the site in 1957.

The quarry's name has turned over a few times: its original designation was Inland Cement, then Genstar, and later it reverted to its old name, Inland Cement.

Looking up from Cadomin's main road, the observer can easily see the quarry atop a resident mountain. However, access to the quarry high above the valley floor is not so simple — the roads leading up are extremely steep and narrow.

In the quarry, limestone is blasted from the mountain, loaded into trucks and then dumped into an inclined hole that extends nearly a thousand feet from surface to an underground crusher.

Cadomin is the only limestone quarry in Alberta that drops its quarried rock down an inclined shaft. Although the dumping method demands that the quarry operate around the shaft hole, the system is so efficient that only 22 men work at the site.

As well, the quarry's efficiency has allowed it to produce between 600,000 and one million tonnes of limestone annually.

Art Bancroft, Shovel Operator

Ron Moss, Quarry Manager

Cecil Vanderveer, Loader Operator

Luscar Coal Mine, Cardinal River

Although the region's coal deposits were discovered in 1912, the original Luscar Coal Mine near Cadomin was founded some time later, in 1921. One of the three largest mines in the Coal Branch area, Luscar Collieries produced 11 million tonnes of coal in its lifetime.

The mine closed in 1965, after losing key markets with the railways and industries. A year later, the large Japanese trading company, Mitsui, approached Luscar Ltd. to reopen the mine to supply coking coal to the Japanese steel industry.

Surface mining operations started in 1968 when a new company, Cardinal River Coals Ltd., was formed as a result of a joint venture between Luscar and Consolidation Coal Company. The initial contract was for the annual shipment of one million tonnes of medium-volatile bituminous coking coal to Japan, via unit trains to Vancouver. Since its opening in 1970, the mine has undergone several expansions, increasing its annual capacity to 2.5 million tonnes.

The mine's main coal seam — the Jewel Seam — averages 10 metres in thickness and expands to 40 metres in some areas. Operations involve simultaneous mining of multiple open pits using the truck/shovel method. Overburden is removed in a series of benches, and is either used to backfill pits or hauled to waste dumps. After mining, the coal is cleaned and dried to specification in the preparation plant facility.

During the early 1940s, Luscar was a town of about 300 miners and more than 650 people in total. The town had all the important facilities for the miners and their families. The town's beer parlour — along with its pool hall — played a major role in miner entertainment. In Luscar, men returning home from the mine had to walk past the bar, and it was a natural instinct for them to stop and restore liquids and salts lost during the day's hard work. Underground mining was known to be dirty, thirst-creating work and, not surprisingly, miners were often hard-drinking men.

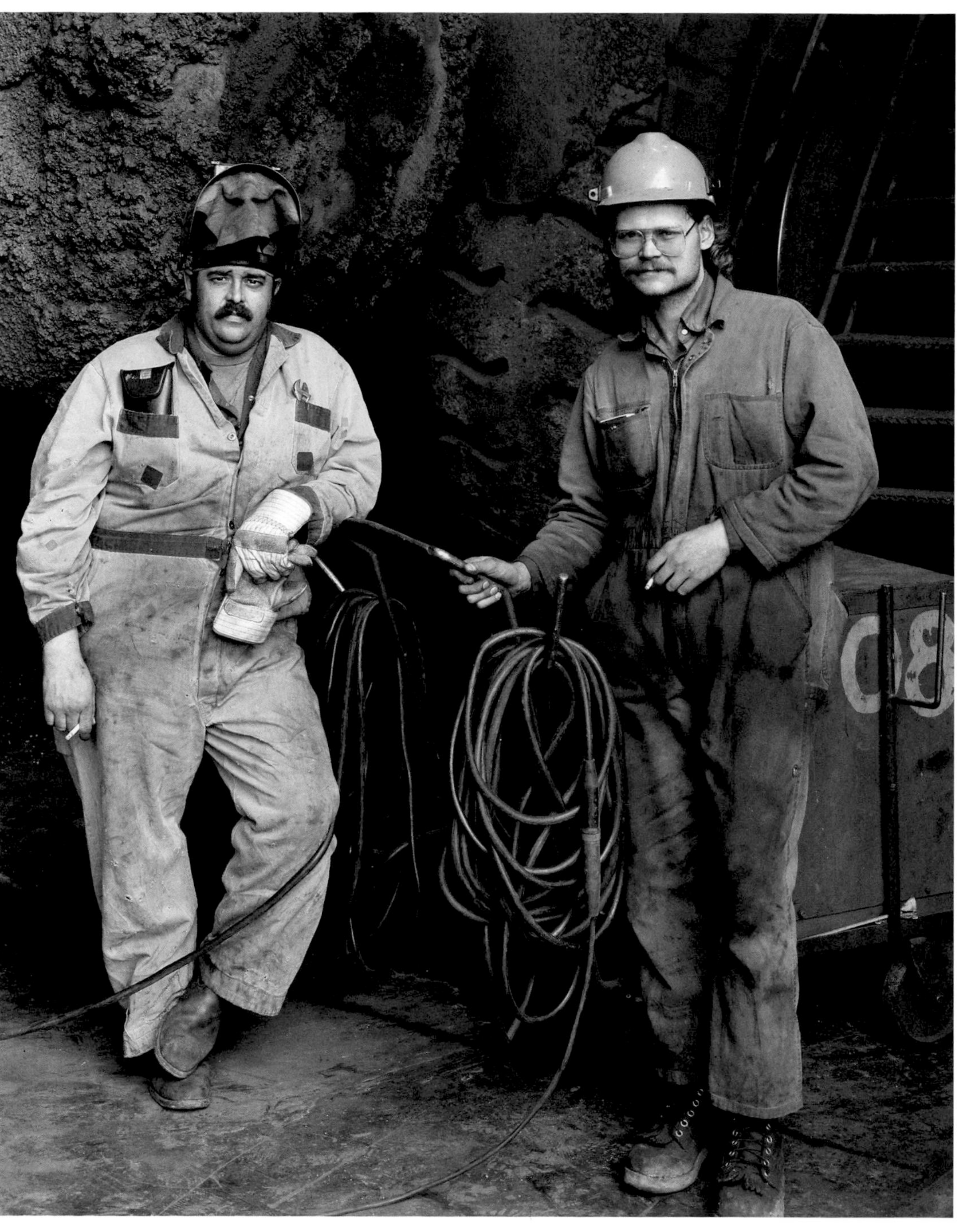

Dean Pfannmuller, Welder and Warren Power, Machinist

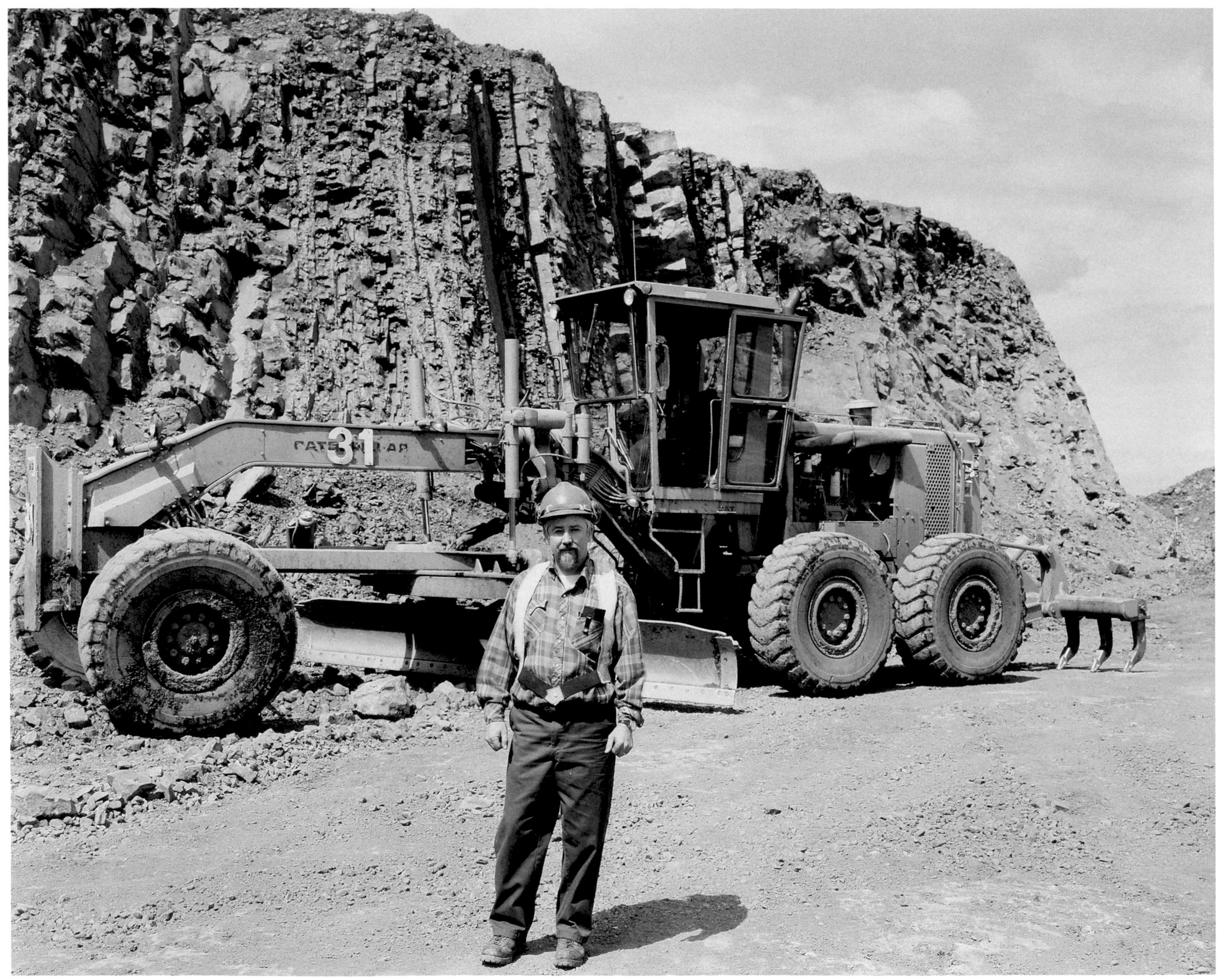

Bert Charbonneau, Grader Operator

Cornelius Bax, Janitor, Bob Porsnuk, Welder, Duane Gubbe, Millwright and Eric Satre, Electrician

Gregg River Coal Mine, Hinton

Named for coal discoverer John Gregg, Gregg River is a metallurgical coal mine located 40 kilometres south of Hinton in the Rocky Mountain foothills. Its predecessor in the region was the underground Kaydee Mines which operated from 1912 until the '30s.

During the late '30s, surface mining began in the Coal Branch area, and an associate company of Manalta Coal began its contract surface mining for Cadomin Coal.

Then, in 1958, Manalta Coal predecessor Alberta Coal was granted an option to the Gregg River property of Cadomin Coal. This option was exercised in 1970. After more than 12 years of detailed study, construction of the Gregg River Mine began in 1981. Late 1982 saw mining operation begin at the site, followed by plant start-up in early 1983.

Manalta subsidiary Gregg River Coal Ltd. has a 60 per cent ownership interest in the mine, with the remaining 40 per cent shared by Japanese steel and trading companies. Gregg River's annual production is approximately two million tonnes of coal. Its main mining interest, the Jewel Seam, has a true thickness of 10 metres with structural thickening reaching 30 metres in some cases.

In the first stage of mining, topsoil and subsoil suitable for reclamation is salvaged. The second stage includes the drilling and blasting of overburden and removal by shovels. Coal is then extracted and loaded into trucks by excavators or front-end loaders, for delivery to the processing plant. A rotary breaker station crushes the coal before its processing in the wash plant with a conventional heavy media bath and flotation process. Finally the dry, clean coal is hauled by unit train to Roberts Bank Port near Vancouver, where it is loaded into ships destined for coke ovens at Japanese steel mills.

Gabe Benedekffy, Loader Operator

Vern Olsen, Driller

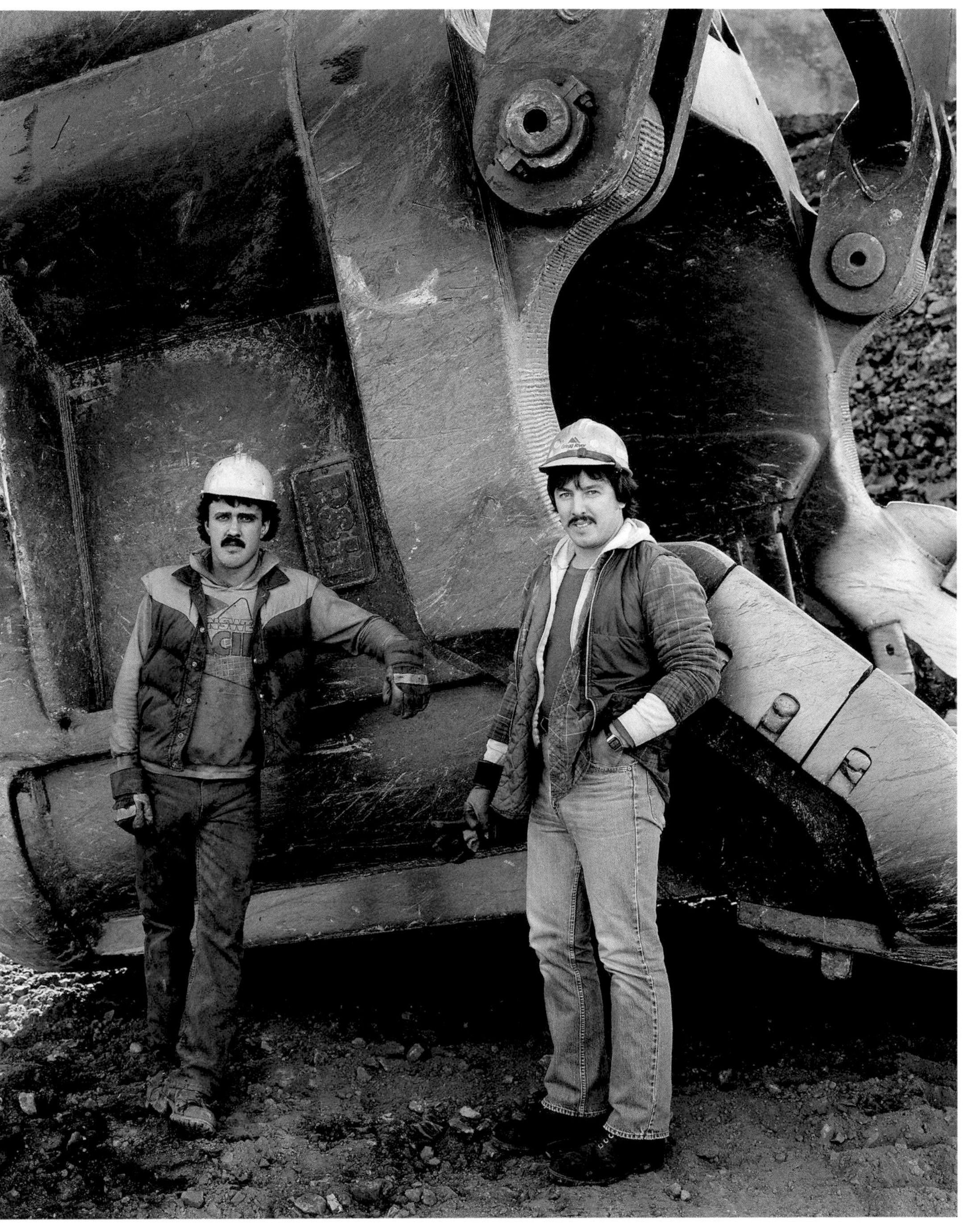

Don Welbourne, Shovel Oiler and Bob Armstrong, Shovel Operator

This was the only small mine in this area. It was built in the early '40s, and was called the Strawberry Creek Mine. Once they went underground with ponies, but they gave that up for some reason.

The old tipple is still here.

I worked here 10 years, stripping coal, before I bought this mine in 1985. Coal sales were fairly good then, but lately we've had such warm winters. Now, I only sell my coal to the domestic market, not to power plants. I sell a fair amount of slack to the County because it's cheaper than sand and salt, and it doesn't hurt the environment. There's about 38 feet of overburden in this mine. I screen the coal in the pit, and then it's stockpiled according to size in sheds. I screen the coal for many customers at four inches and bigger. The coal is higher quality in heat content than most coals around here.

A small guy just can't compete. I can't compete with big companies that sell their coal for $18 or $20 a tonne. I have to sell stove coal for $40 and stoker for $32 a tonne.

Most of the time, I'm a one-man operation, although I sometimes get a bit of help in the fall.

I don't mind this business at all, but there's no money in it. That's the problem. It would be a lot more enjoyable if I was making money at it.

Charles MacDonald, Operator

Genesee Coal Mine, Genesee

In 1980, Fording Coal Limited and Edmonton Power (of the City of Edmonton) formed a joint venture to develop the Genesee Mine.

Located 85 kilometres southwest of Edmonton, the facility was designed to supply thermal coal to the Genesee Generating Station, operated by Edmonton Power. Fording designed, developed and now operates the mine.

Opened in 1988, the mine currently produces 1.5 million tonnes of subbituminous coal annually for the generation of power. Edmonton Power's first 400-megawatt generator started commercial operation in 1989 and a second 400MW unit will be commissioned in 1994. Two additional 400MW units will be incorporated as the demand for electrical power in Alberta grows.

At Genesee, the first step in mining is to recover all existing topsoil and replace it on mined-out areas that have been recontoured.

Second, operators remove overburden with a large electric-powered dragline. An electric shovel loads the coal onto 136-tonne dump trucks for direct delivery to the power station. Three major coal seams are recoverable at this stage.

The final step in Genesee mining is the reclamation of land to a capability equal or greater than before mining began. In fact, reclamation research into efficient methods began in 1981, seven years before Genesee mining activities started. A committee that includes local community members oversees Genesee's reclamation efforts.

When the Genesee mine expands to meet the needs of the second generating unit, approximately 90 people will work at the site. Since production began, the Genesee has received several awards for its outstanding safety record.

Fred Stachnik, Truck Driver and Randy Davis, Dozer Operator

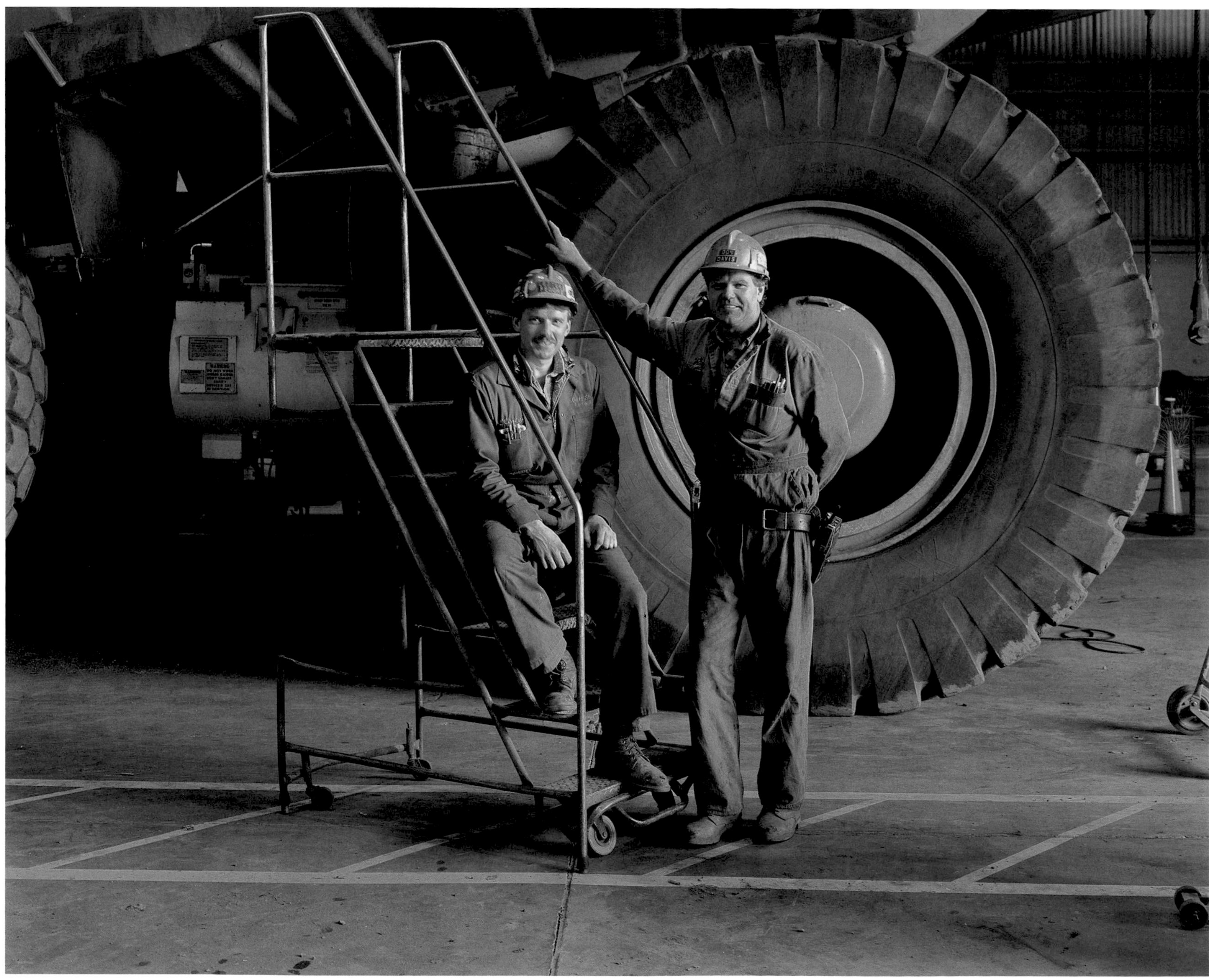

Lestor Fedor and Don Davis, Electricians

Willy Schmidt and Carl Renaud, Mechanics

Highvale Coal Mine, Seba Beach

With annual production of subbituminous coal approaching 12 million tonnes, the Highvale Mine is the largest coal mine in Canada.

Mine owner TransAlta Utilities owns all the mine's major equipment and facilities, as well as the mine property near Seba Beach, south of Wabamun and west of Edmonton. Manalta Coal Ltd. is the facility's mining contractor.

In the early days, local farmers mined the shallow coal themselves. Commercial surface mining by Mount Royal and Continental Collieries began along the south shore of Lake Wabamun between 1943 and 1961. Then in the late '50s and early '60s, an exploration drilling program recommended the construction of a large mine-mouth power station in the area. When the first unit of the Sundance Station was completed in 1970, the Highvale Mine began production.

As additional electrical power generators were added to the facility, the demand for coal from the Highvale mine increased. In the early '80s, the Keephills power station was constructed next to Highvale, necessitating mine expansion from eight to more than 11 million tonnes of coal annually.

A mine of this size requires modern surface mining equipment. The most prominent machines at the site are four walking draglines for overburden stripping. Recovery of coal from the multi-seams is further accomplished with shovels, loaders and a massive fleet of 150-tonne trucks.

Miners at the Highvale Mine generally come from the neighbouring farming communities. For younger workers, a high school diploma and comprehensive on-the-job training is essential to enter a career in mining.

Ed Fortier, Dragline Operator and Doug Moger, Dragline Oiler

Every summer, surface mine rescue teams representing various companies gather for the Alberta Surface Mine Rescue Competition. Sponsored by the Alberta Mine Safety Association, the event is held in a different location every year, and gives the trained teams the opportunity to demonstrate their skills in a new and untested environment.

Each team consists of six members and a spare; one member acts as a team captain. The group generally selects its members from a variety of company functions, with office workers, mechanics and supervisors joining forces. Behind the scenes, each team consults trainers and advisors.

The competition aims to provide rescue teams with a variety of different emergency situations that might be encountered on the job. Each team responds to calls for first-aid treatment, search and rescue in a smoke-filled building, and fire control. A simulated accident at a mine site would commonly involve mine equipment and severe injuries to miners.

The surface mine rescue teams are primarily trained for accidents at their own sites, but in the event of a major accident afflicting neighbouring publics, their respective companies would sponsor their participation in a larger rescue effort.

Surface Mine Rescue Team - Back row: Sergei Ewachniuk, David Richardson, Byron Karesa, Karl Kusmierz
Front row: Stan Worms, Larry Eberle, Captain and Chris Mastowich

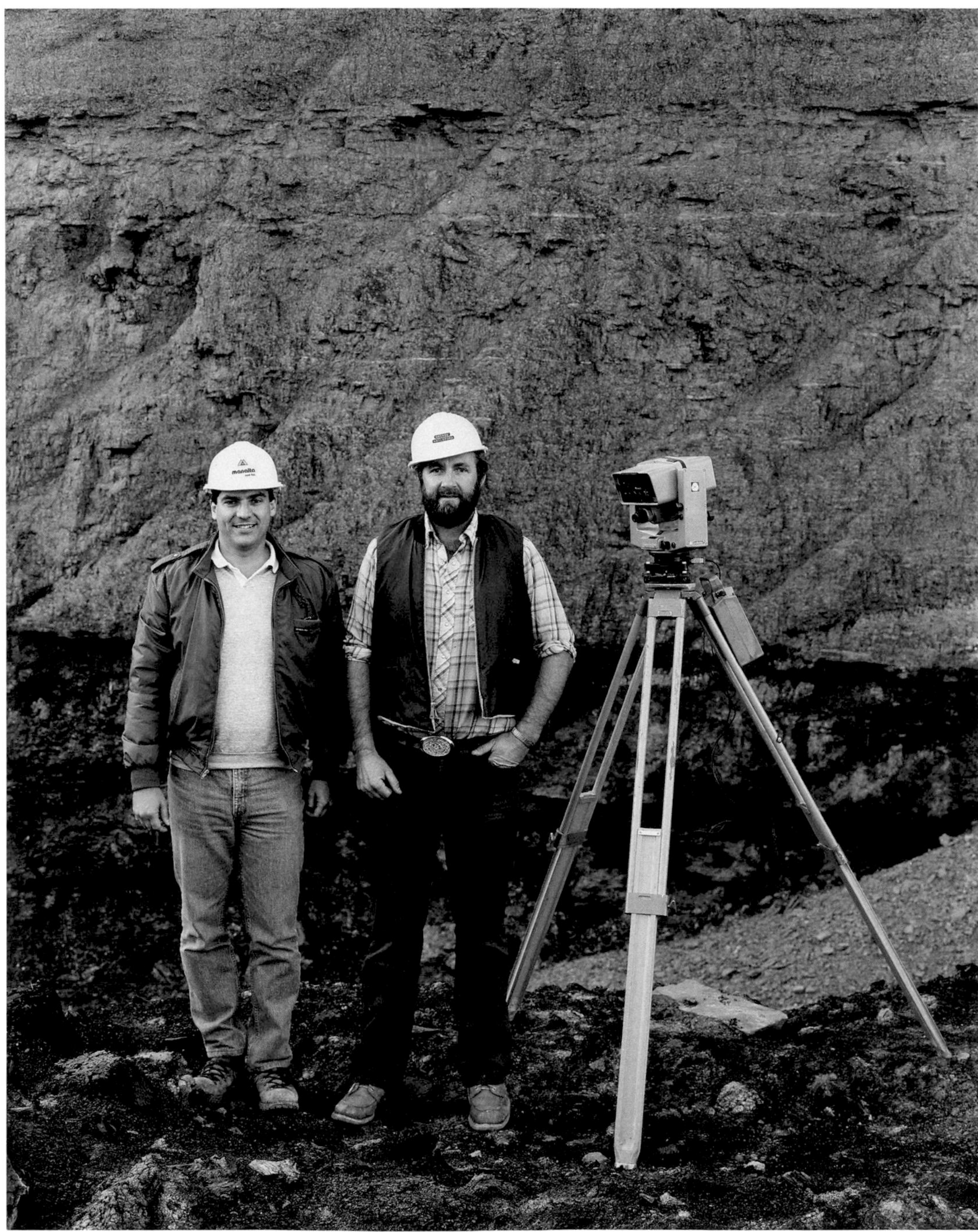

Joe Derosier, Mine Engineer and George Astleford, Surveyor

Mines/Companies

Smoky River Coal Mine,
Smoky River Coal Ltd., Grande Cache

Obed Coal Mine,
Luscar Ltd., Hinton

Whitewood Coal Mine,
Transalta Utilities Corporation/Fording Coal Limited, Wabamun

Egg Lake Coal Mine,
Egg Lake Coal Co. Ltd., Morinville

Consolidated Sand Pit,
Consolidated Concrete Limited, Villeneuve

Sil Silica Sand Quarry,
Sil Silica Inc., Bruderheim

Thorhild Coal Mine,
North Point Coal Co. Ltd., Thorhild

Canadian Salt Mine,
The Canadian Salt Company Limited, Lindberg

Suncor Oil Sands Mine,
Suncor Inc., Fort McMurray

Syncrude Oil Sands Mine,
Syncrude Canada Ltd., Fort McMurray

Underground Test Facility,
Norwest Mine Services Ltd./AOSTRA,
Fort McMurray

Northern Alberta

Locations of photographs and mines

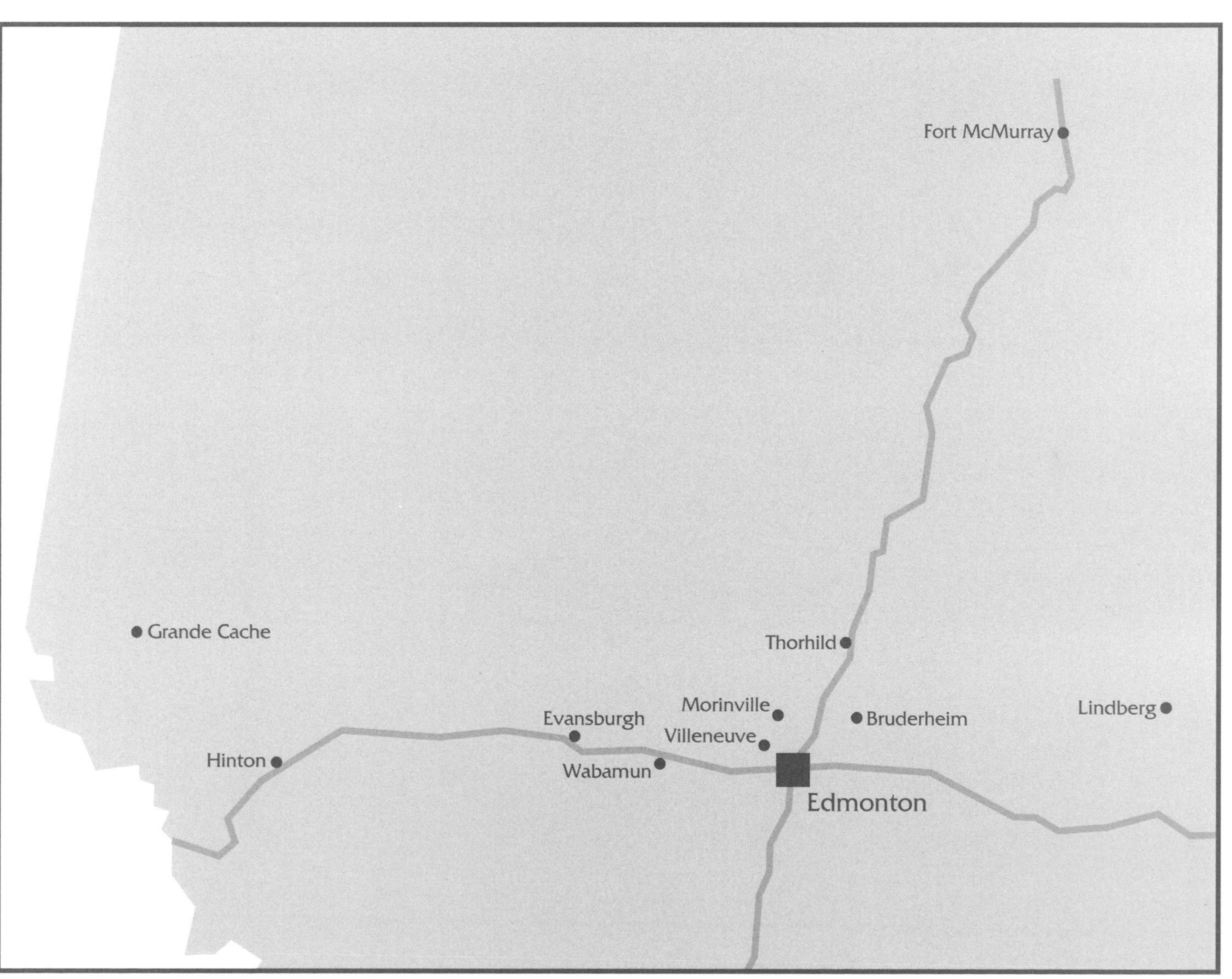

Smoky River Coal Mine, Grande Cache

Coal exploration began in the Smoky River area as early as 1926, but it was not until the mid-1960s that serious commercial development of the industry began.

Stakeholders decided in 1968 to proceed with coking coal production for the expanding Japanese steel industry, and the Smoky River Mine was developed to produce two million tonnes of premium grade coking coal annually for 15 years. Longwall mining, an underground method using high-speed cutting machines and roof support equipment along a long wall of coal, was used in this progressive Alberta mine.

Initially, Smoky River opened two underground mines. As reserves in the initial mines became depleted, new underground and surface mines were developed to maintain productive capacity. Today, only 45 per cent of total production comes from underground mining, with the remainder mined by conventional mountain mining truck and shovel methods.

Two significant factors contributed to Smoky River's success: one was the establishment of the town of Grande Cache, a beautiful and remote site where the company built miner homes and other facilities. On the marketing side, the signing of a long-term contract with the Japanese, plus the establishment of an adjacent power plant to burn tailings, gave the mine a healthy head start in the marketplace.

Coal miners from across Canada and around the world were attracted to the Smoky River Mine. In fact, many underground miners from operating Alberta mines have at one time worked at Smoky River.

Cape Bretoners Lloyd, Bobby and Malcolm came as experienced coal miners to Grande Cache from an uncertain Nova Scotia workplace. Each of them have been in this mining town close to 25 years, but will probably return to Cape Breton Island when they retire.

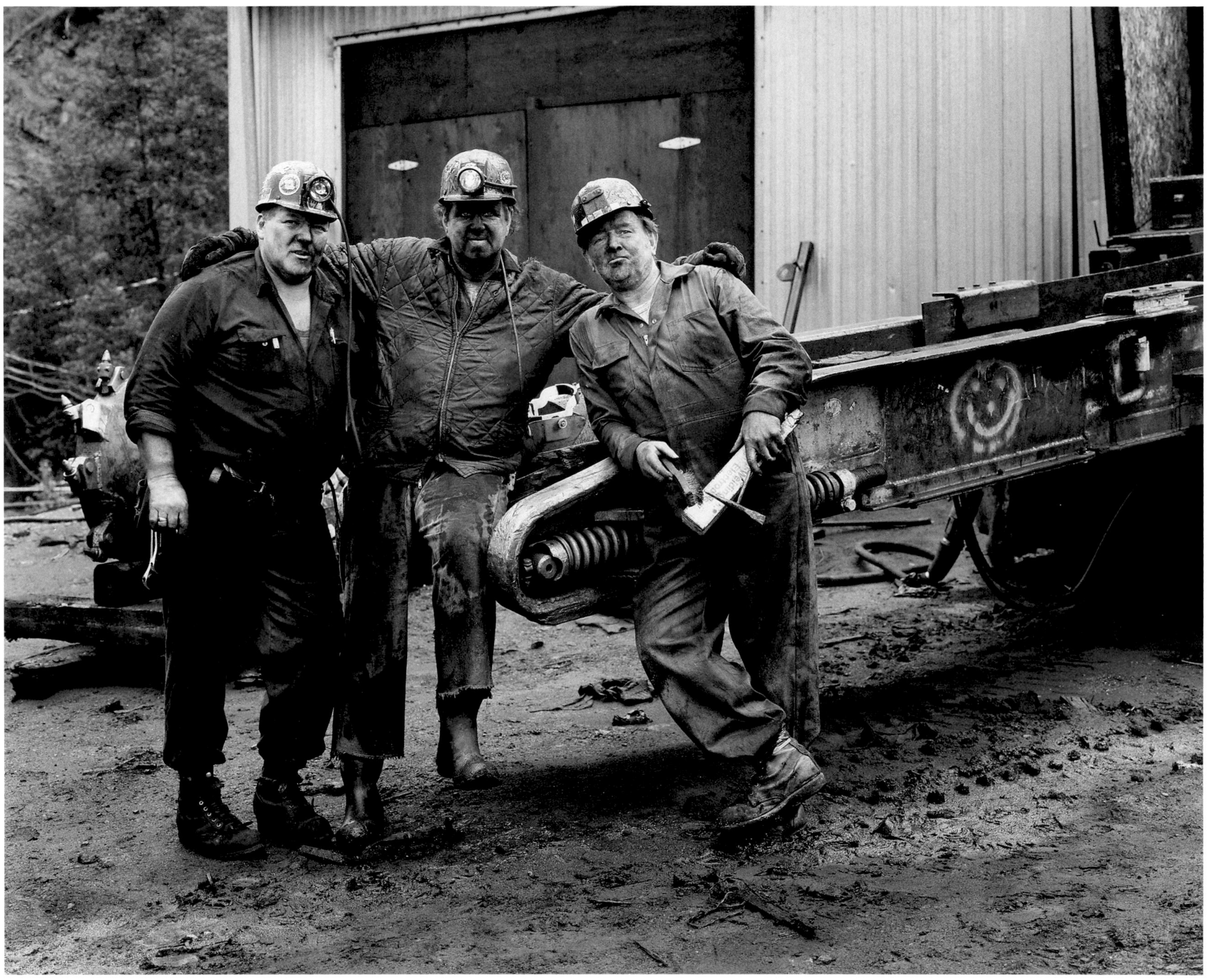

Lloyd Layes, Bobby Stewart and Malcolm McNeil, Underground Mechanics

Gerard McNeil, Lloyd Lewis and Norm Johnson, Underground Foremen

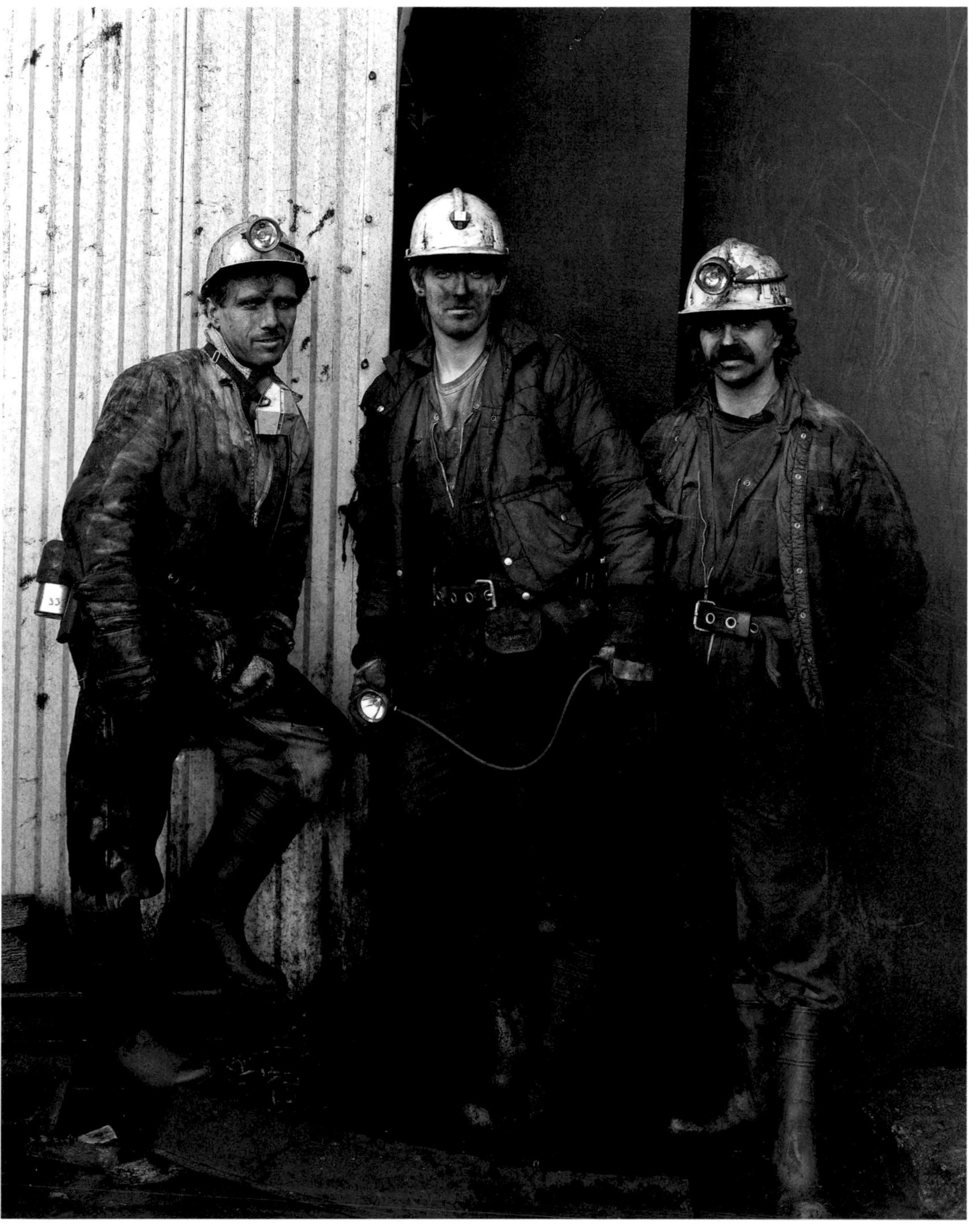

Rick Harvey, Tim Sandkuhler and Jerzey Tepek, Underground Miners

Rob Moore, Mechanic, Dean Winsor, Warehouseman and Ron Mortensen, Foreman

Jim Petrie and Terry Butterfield, Welders

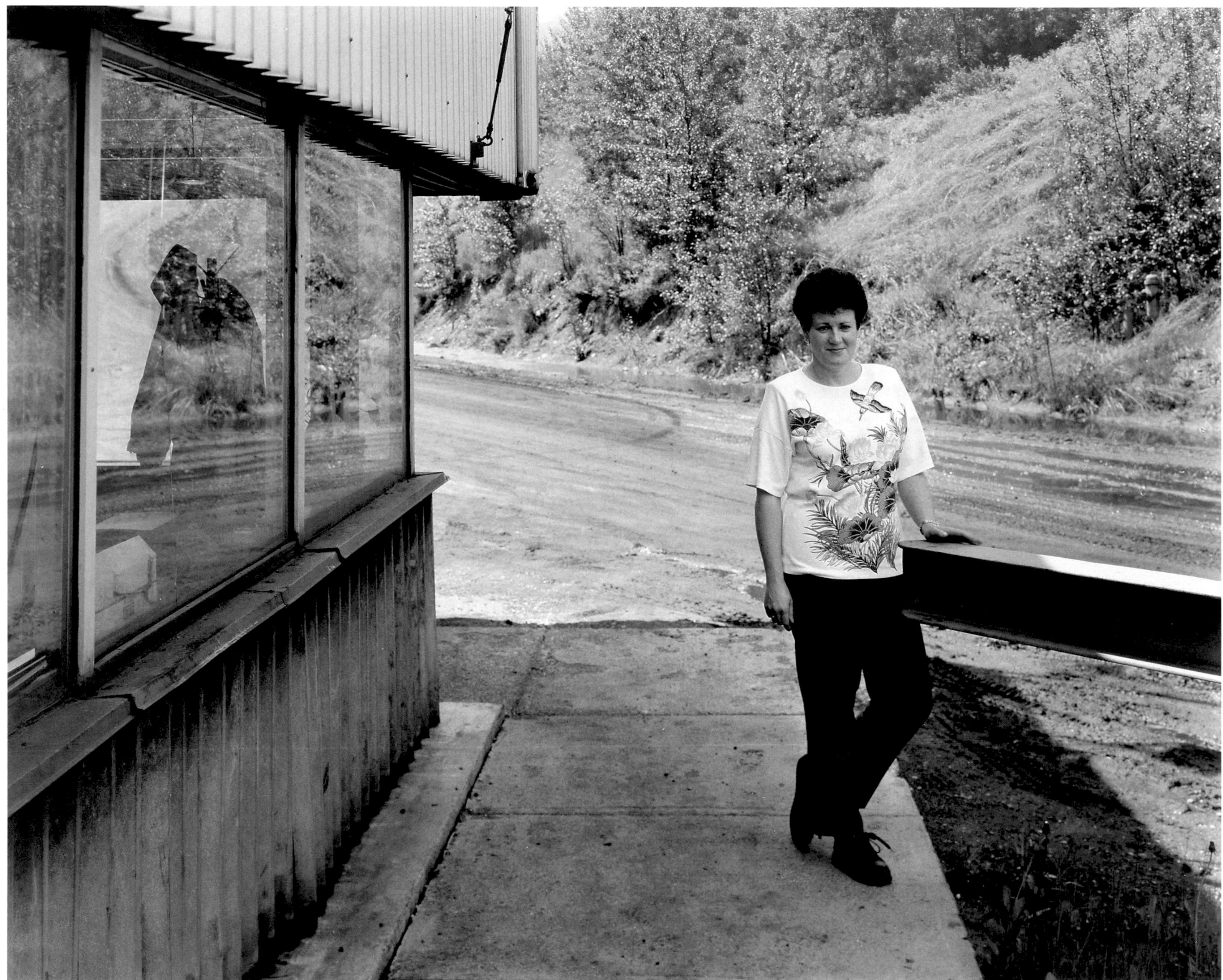

Marylin Flaumitsch, Security Guard

Penny Johnston, Sampling Technician

Obed Mountain Coal Mine, Hinton

The Obed Mountain Coal mine is located 25 kilometres northeast of the town of Hinton, in the Rocky Mountain foothills. Construction of the mine began in 1981, and Union Oil officially opened the facility in 1983. In 1989, Luscar Ltd. purchased the Obed Mountain mine.

Although there are no records of early coal mining at Obed Mountain, considerable mining did take place nearby in the small villages of Hinton, Pocahontas and Brule along the Athabasca River Valley. The renowned Coal Branch towns were situated not far south.

Coal at the Obed site is surface mined, with overburden removed using the 67-cubic-metre "Athabasca Rose" walking dragline. Mobile equipment including dozers, scrapers, shovels and trucks support the dragline operations. Trucks with 108-tonne capacities haul the coal to a preparation plant where a jig process cleans the coal. The cleaned and dried product is then loaded onto an overland conveyor that transports it 10 kilometres to load-out facilities on CN Rail's main line.

The Obed mine produces high volatile bituminous coal for electric power stations in Pacific Rim countries. Coal is shipped 960 kilometres by 100-car units to the large coal port of Roberts Bank, British Columbia.

Current production is estimated at one million tonnes annually.

Norm Jewett, Dozer Operator and Bob Peeke, Team Leader

Kim McCallum, Dragline Oiler

Kevin Wallace, Preparation Plant Technician and Morley Shaw, Team Leader Preparation Plant

Charles

I was born in Alberta in 1910. When I was 15, my dad was hurt in the mine, so I had to make a living for the family.

I started in the Evansburg Mine, picking rocks on the picking table. Then I helped lay track underground. After that, I was a timberman's helper with my dad. Then I got my papers, and went digging coal. Evansburg closed down in 1936, so I moved to the Coal Branch and worked on the Mountain Park tipple.

I felt right at home underground. Although I wasn't a very big guy, I was really tough — tough as long as I lasted!

After that, I had my own mine, the Pembina Peerless Coal Company, down the Pembina River. I started it all on my own. I sunk in the little money I'd saved at the Coal Branch, and had to work harder than ever.

Then I worked in surface mines for 28 years, until 1973 when I retired. I'm not going to dig any more coal at my age. I've had enough of the coal!

And coal doesn't look black to me anymore — it looks kind of brown!

Charles Ostertag, Retired Underground Coal Miner and Surface Mine Operator

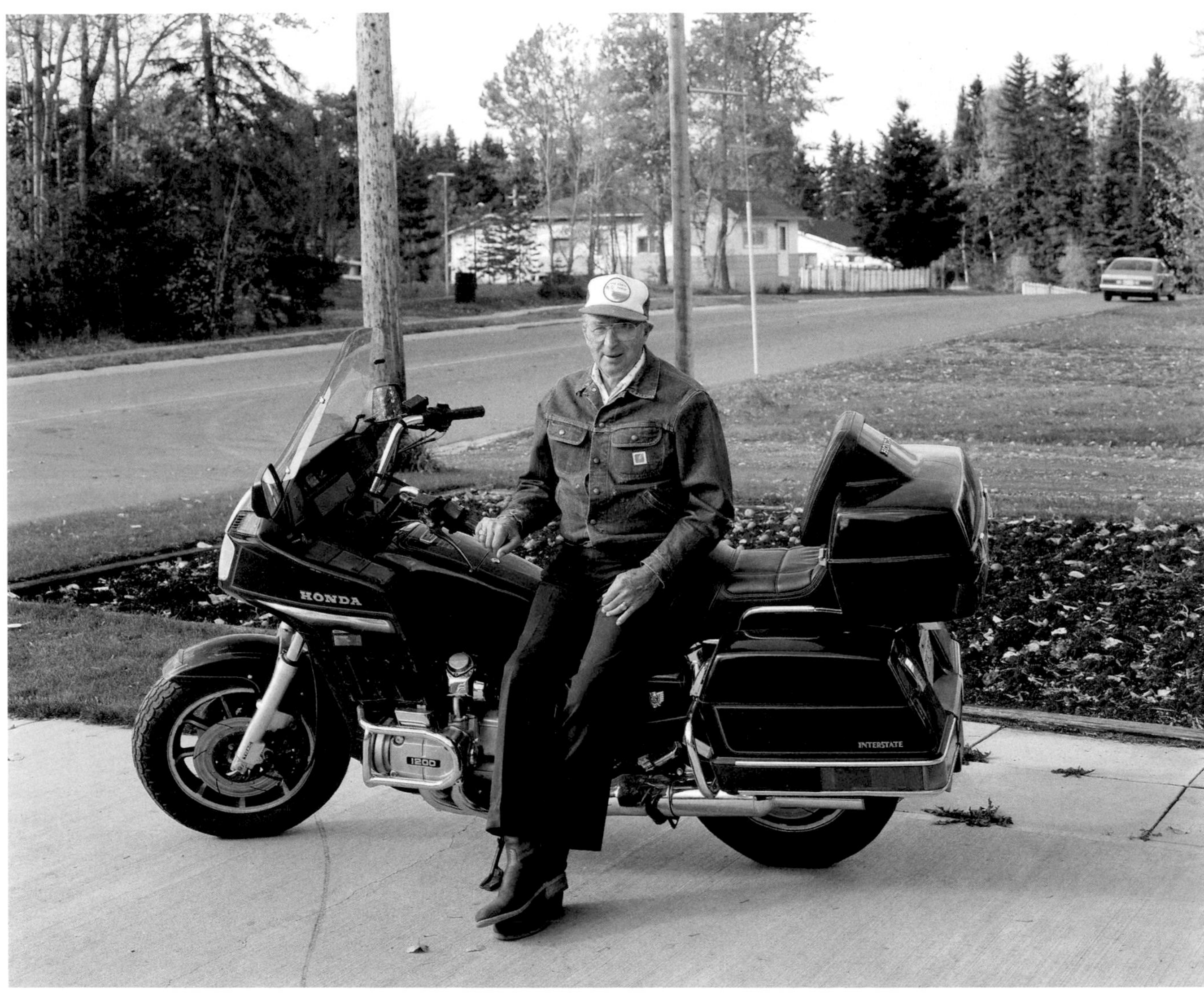

Fred Woollard, Retired Surface Coal Miner

Fred I bought a motorcycle when I was 66 years old. It weighs 750 pounds. You have to respect something like that because it's so powerful — it's not a thing to take lightly. It's just something to get some satisfaction from.

Both my dad and grandfather were fishermen, and they settled here between the Lac St. Anne and Wabamun lakes. My twin brother and I were born in Wabamun in 1919.

At the age of 25, I worked underground in the Lakeside Mine as a driver with horses. It was handy and a place to get a job, so I took it. Then I worked at the Atlas Mine in East Coulee, and I was there during the flood. When the underground mines shut down, I got a job here in the strip mines.

I started as a tippleman, then over the years, I got on the cats and loaders. Eventually, I got on the big dragline brought in from England. Yeah, I was fortunate enough to be the first operator they trained on that machine, and I worked on it for nine years. I liked being around heavy equipment. And it was a prestige job to operate the dragline. At the time, it was the largest of its kind in Canada. But after a while I got tired of night shift because it buggers up your social life.

I worked 34 years in the Whitewood Mine, plus six years underground. I always liked underground mining, but when I quit, I never had any hankering to go back.

Tom My nickname is Eagle, because when I was young, I used to climb trees. I was always on the top of a tree, like an eagle. Jesus, when I was young, there was nothing but trees.

My folks homesteaded in Beaufort, 15 miles west of Leduc, from 1897. I was born there in 1911. When I was 31, I went from the farm to Robb and got a job in the mine. I was promoted to the shop where I did everything from welding to repairs, both underground and on the surface. I was the last man in the Robb Mine before they flooded her.

From Coal Valley, I moved to Wabamun where I helped build the tipple. I bought my own truck and delivered coal to northern Alberta from the underground and later the surface mines. I really enjoyed the trucking.

Tom Babiak, Retired Coal Truck Driver and Friends

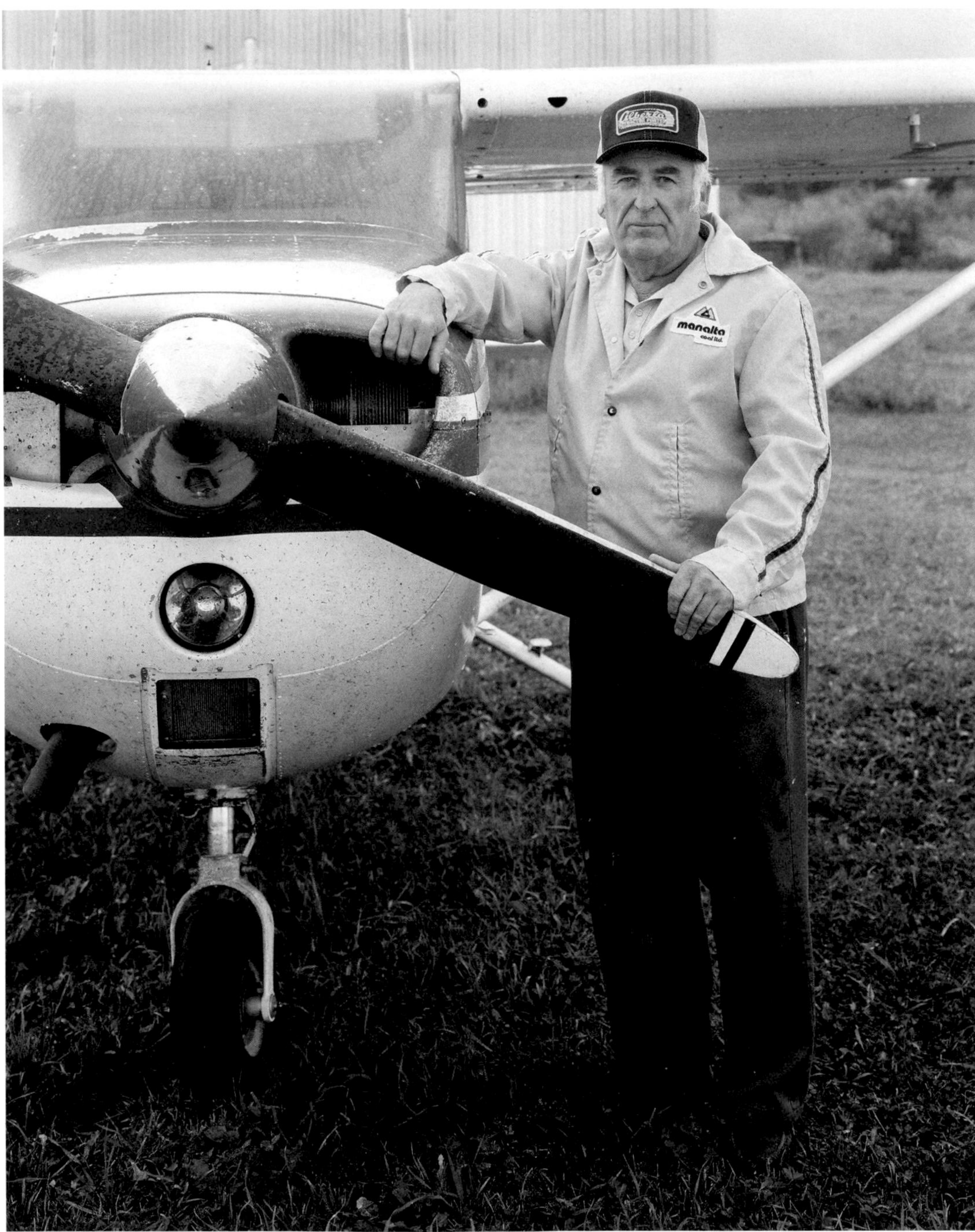

Louis Carriere, Retired Coal Mine Manager

Louis

I was born in Legal, Alberta in 1926. I've been in the coal mines all my life, some forty-odd years.

In 1944, when I was 19, I was a cat operator stripping coal at Cadomin with Mannix Construction. I became a shovel operator within six months. I was there until 1948. Then I went to Mountain Park. When that mine shut down, I loaded the equipment out of there. I then moved to Forestburg, and operated the Page 8-yard dragline, which was at that time the biggest machine in the country. After that, I worked in the Estevan and Wabamun areas for Manalta Coal.

Equipment always used to fascinate me. Even as a mine manager, I used to go out to the pit every day, load a little coal, and run the shovels and draglines. I think the people I worked with appreciated that a mine manager could run the equipment. They'd say, "Hell, you can't do this, you can't do that." I'd say, "Get off the seat, and I'll show you!"

I've got my own airplane, a Cessna 172. And I still play hockey. When I was 16, I played senior hockey — oh, we played hockey in those days. Lately, I've played in California with the Edmonton Oldtimers, all old pro hockey players. Of 38 teams, we finished second.

Oh, I always played hockey.

Located in northern Alberta just north of Lake Wabamun and west of Edmonton, the Whitewood Mine is one of the earliest large surface coal mines in the province. TransAlta Utilities owns the mine's reserves and all equipment, but contracts the coal mining to Fording Coal Limited. The Whitewood plant was established to feed the Wabamun Generation Station. Originally, the station was gas fired but eventually was converted to a mine-mouth, coal-fired power station.

The Wabamun area's rich mining history begins in 1910, when the first mines started at the outcrop and went underground along the usually flat-lying seams. A total of five thick seams of subbituminous coal occurs in the area. Underground mining continued at a few mines until 1948, when the first surface mine in the area began stripping overburden with the use of horses. Tractors inevitably replaced the horses at these operations.

During the early 1950s, TransAlta studies indicated that Alberta's large coal reserves were the optimal fuel for future electricity generation, and that the Wabamun station operations should be converted from natural gas to coal. So in late 1962, mining commenced at the Whitewood, with an annual production rate of approximately two million tonnes.

The mine uses a dragline for overburden removal, with front-end loaders and electric shovels recovering the raw coal for trucking to the tipple. The run-of-mine coal is crushed before it is conveyed to the Wabamun Generation Station.

Retired miners in Wabamun remember the days of underground mining as occurring within the town's current boundaries, and recollect the primitive stripping methods first used when surface mining began.

When the 30-yard Ransomes-Rapier dragline was first erected at Wabamun, it quickly became a tourist attraction for locals and curious Albertans.

David Denton, Shovel Operator

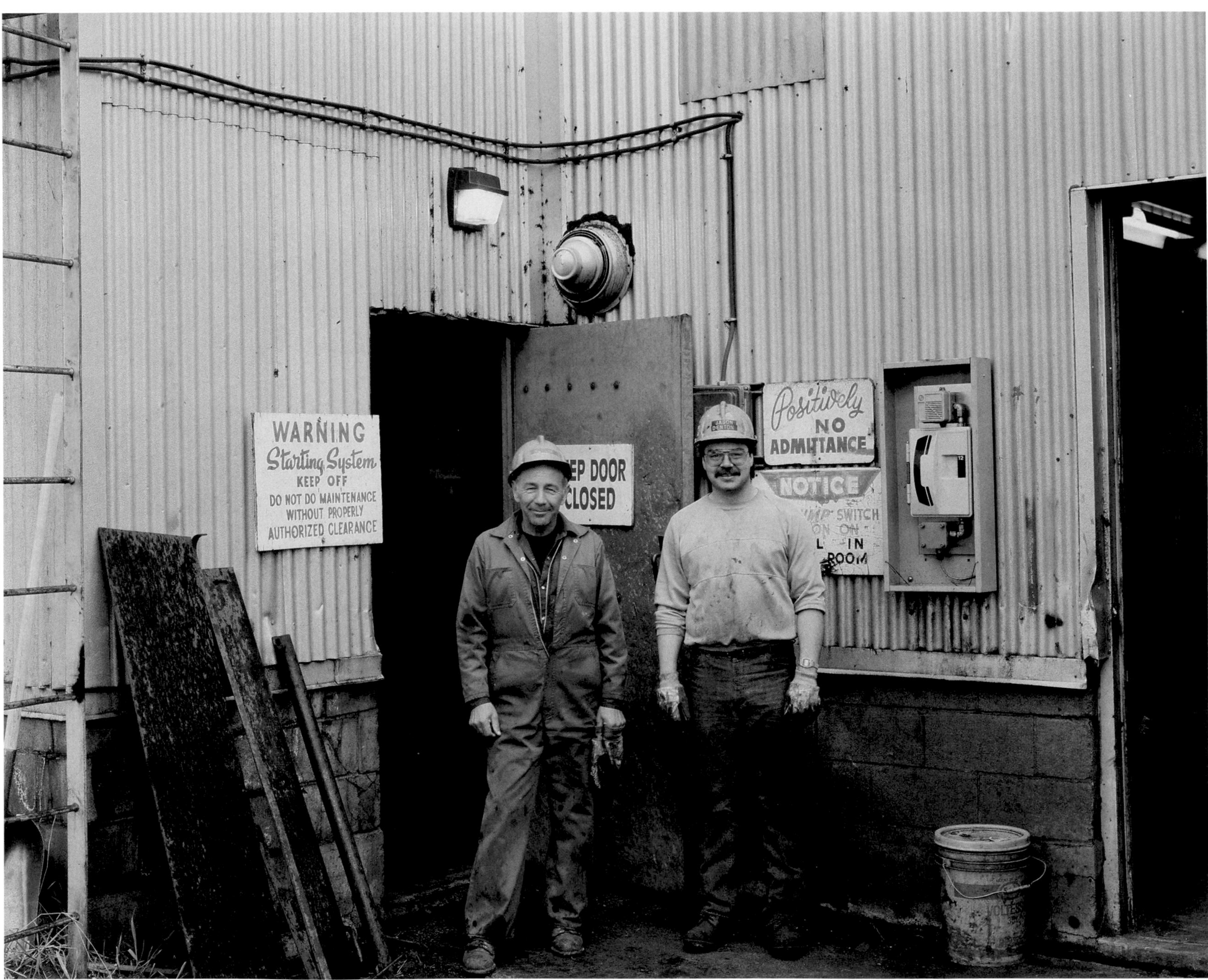

Willy Wichuk and Jason Denton, Transfer Tipplemen

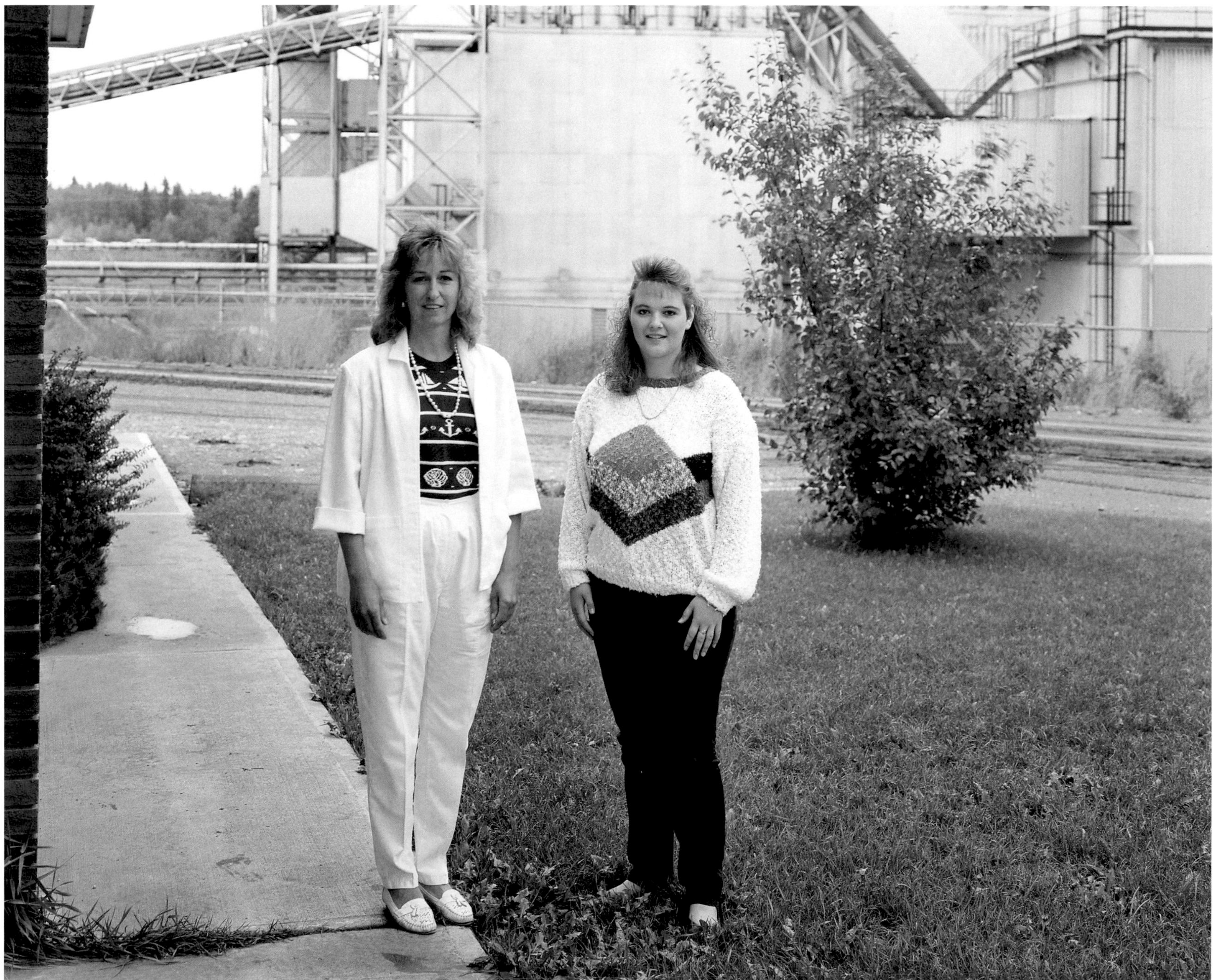

Maureen Colban, Secretary and Chris Williams, Summer Student

Egg Lake Coal Mine, Morinville

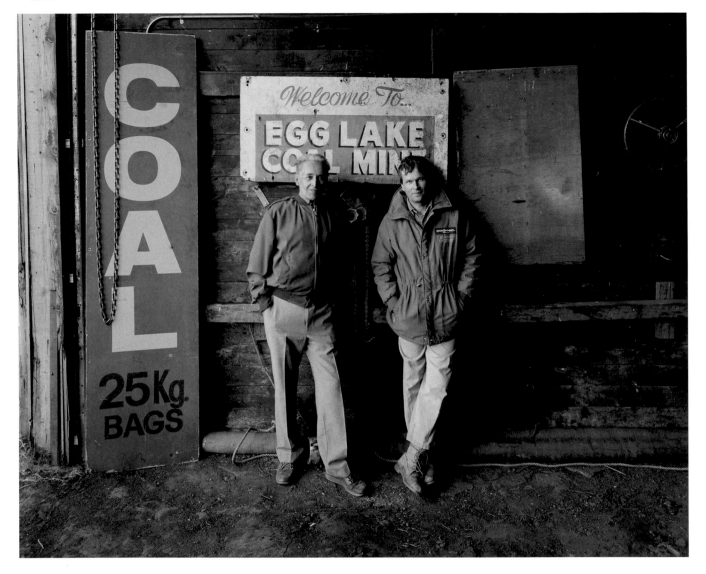

Tom I was born in 1915, about four miles northwest of here.

My granddad squatted here sometime between 1885 and 1890, we're not sure exactly when. The first mine here was developed on the surface in about 1915 — they discovered coal in a water well about 10 feet deep.

When my granddad died, my father moved onto the farm, but he didn't do any mining. In 1940, I bought the mine and the farm from the estate, and a year later, started Egg Lake Coal with the help of an experienced miner from the Coal Branch.

At first, we used a horse to dump the coal in the tipple, then later we bought an engine. We had to blast the eight-foot seam and then pick up the coal with hand shovels. We moved the Egg Lake School here to use as a bunkhouse for men in the old days — until then, we'd been using a boxcar. Up until the early '60s, there were 20 people working and living here.

In the '50s, I'd had enough of the business, and couldn't even look at a piece of coal any more.

I love coal mining. I was born 40 years too late.

In Tom's day, everybody burned coal. The horses and wagons were lined right out to the road. Slack coal was sold to the Edmonton power plant on the river, the one near the Low Level Bridge.

The strip mines got a real boost when all the underground mines went on strike. Now the big customers are the poultry and hog farmers.

Tom Logan, Retired Coal Mine Operator and Chris Hart, Manager

The Consolidated sand and gravel quarry at Villeneuve opened in the early '60s as a small operation and has grown to become the largest washed-aggregate producer in Alberta. Sand and gravel quarrying goes back so many years in the Villeneuve area that working in the sand pits has become a venerable family tradition.

In 1967, Consolidated began its large-scale operation at Villeneuve Pit #45. Because a high water table exists in the swampy area, the entire quarry procedure starts with a dragline, which digs the sand and gravel deposit. A conveyor system moves the run-of-mine material to an open-air crushing, washing and sizing complex. At this stage, a recovery unit investigates the product for finely scattered gold that occurs in small amounts throughout the gravel.

Because the wet sands and gravels at Villeneuve freeze in the winter and make mining extremely difficult if not impossible, quarrying at this site is a seasonal operation. In winter, only 10 employees work on maintenance, whereas in summer as many as 60 people are involved in operations. During peak periods of construction activity, it is not uncommon to see a long stream of highway dump trucks lining the road to the Consolidated pit.

Commonly known as aggregates, sand and gravels are used in ready-mix concrete and asphalts, and range from extremely fine to coarse sizes. Careful sizing gradation is important in the business, as every market has its own unique needs for the product.

Joe Dubovsky, Superintendent

Consolidated Sand Pit, Villeneuve

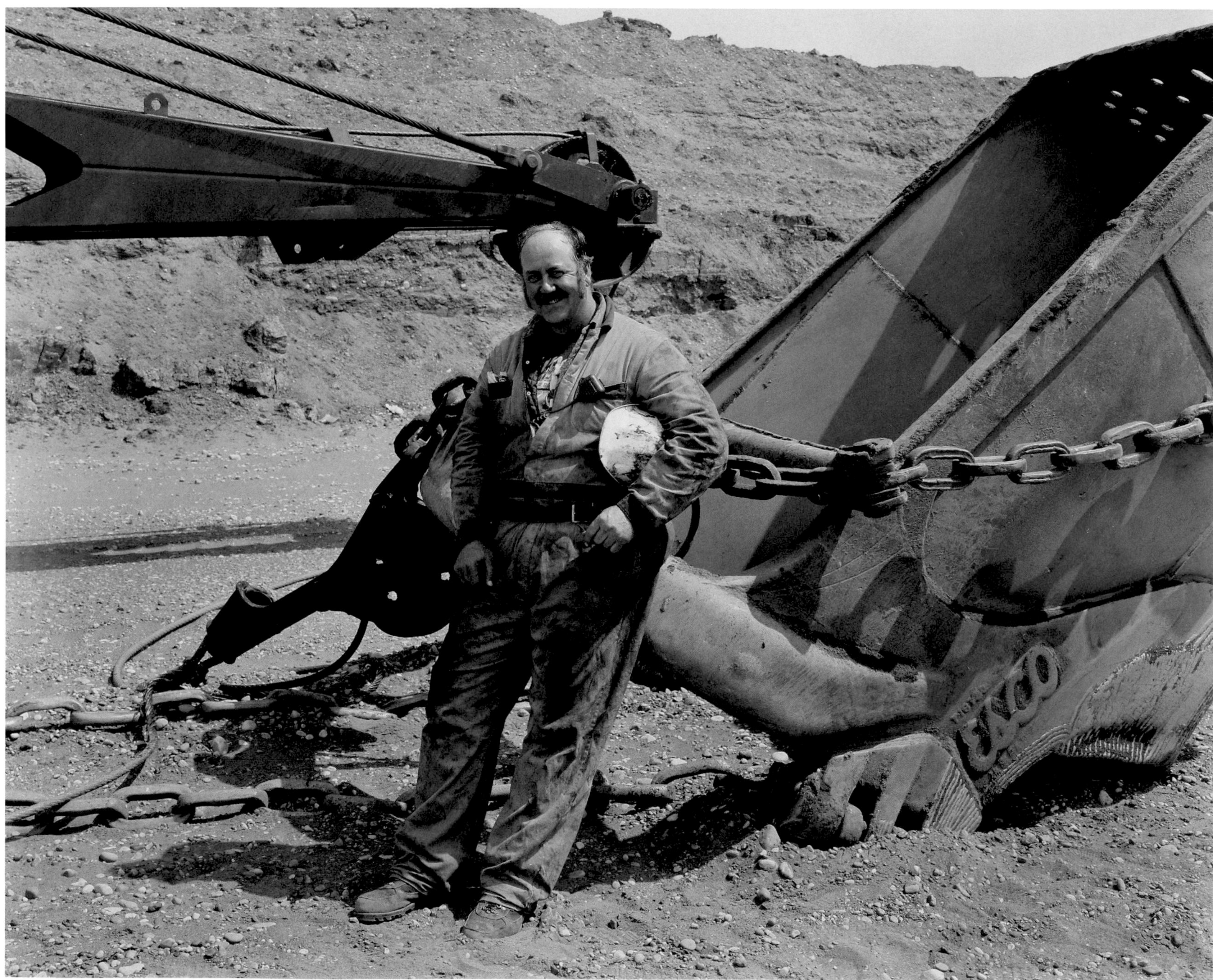

Richard Edwards, Dragline Oiler

Duncan When I first came to Alberta, the mining industry was all coal men. Now there are no coal men left.

I was born in 1903 in Newcastle-Upon-Tyne. I studied mining engineering, but there were no jobs for engineers, so I immigrated to Bienfait, Saskatchewan in 1928. From there I moved to Mountain Park, Alberta, where I worked as a miner, a ventilation man, and a fireboss. After eight years, I got kicked out for trying to form a union for fireboss-es. In the early '40s, I became a pit boss at the Saunders Creek Mine in the Nordegg area. It was the only time the mine made money.

I also worked at the Nordegg and Alexo mines. When I left Alexo, I became Mines Inspector in the Drumheller area. My job was to check the mine's ventilation and working places, according to the specifications of the Coal Mine Act. I was responsible for 57 operating mines, including 21 in the Drumheller Valley.

When I was inspecting the big mines in the Pass, I would go underground in the morning with the miners and the mine manager. Between the manager and me, we would have only an orange or a candy bar to last us all day in the mine. We walked all day throughout the dirtiest parts of the mine. The next day, I'd do the same thing at another mine. My biggest concern was always the detection of methane gas.

After retiring, I did some prospecting for coal, but my hobbies are painting pictures and feeding the blue jays every day.

Running a mine had its peculiarities. The problem was that the miners never looked upon changes as beneficial. All coal miners had an innate suspicion of the mine manager.

Duncan Brown, Retired Chief Mine Inspector

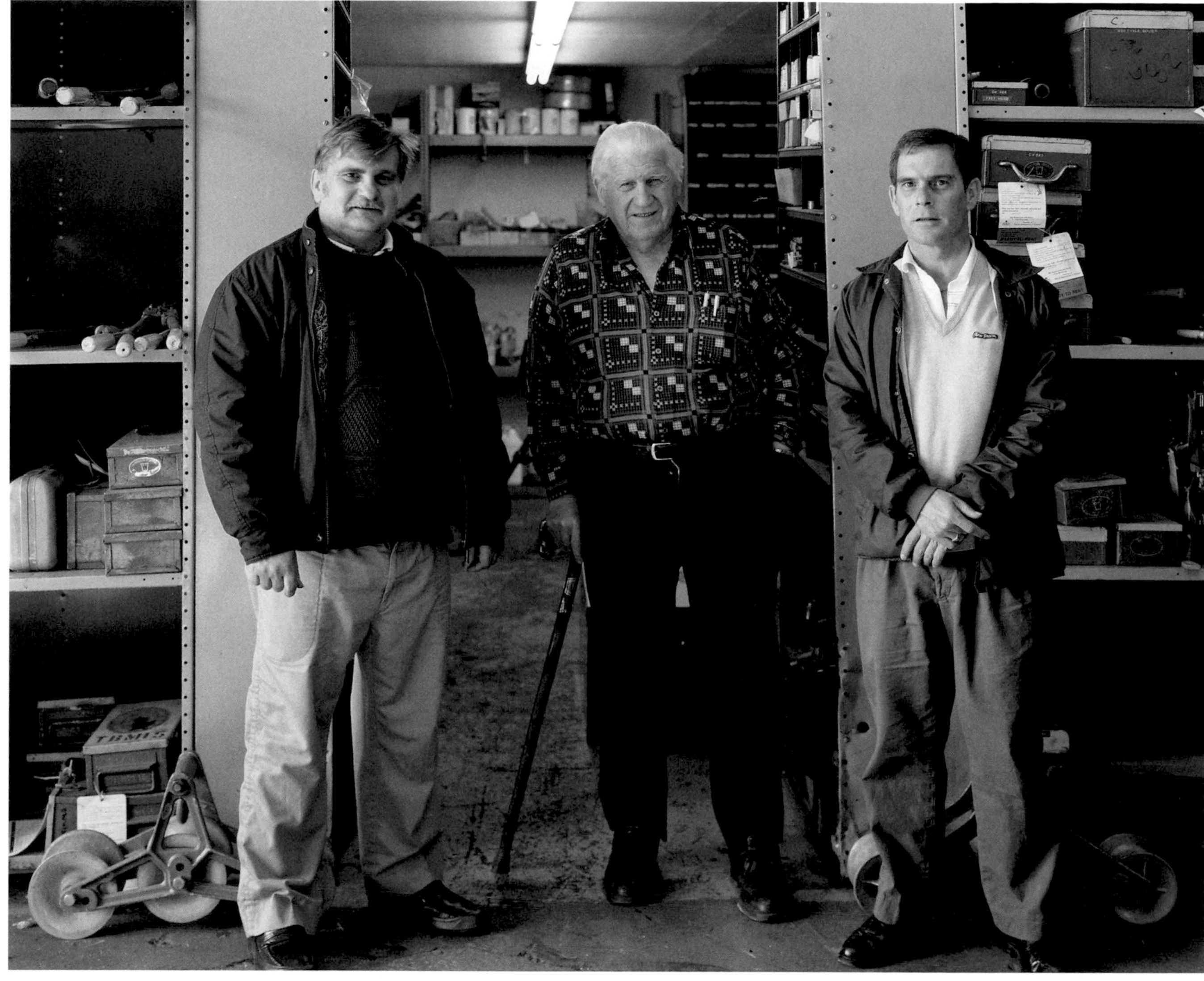

Richard, Ernie and George Wirtanen, Owners

"The man who once most wisely said,
Be sure you're right, then go ahead,
Might well have added this, to wit:
Be sure you're wrong before you quit."

from Steinmetz

Ernie My dad, who was born in Finland, was a first-class steam engineer in the mining and lumbering industry. I was born in Dollarton, British Columbia in 1919. After I finished school in 1936, my dad gave me a job at Britannia Copper Mines to take out a 1000-kilowatt turbine. When I'd worked a few years, I went back to school and studied electrical engineering.

My first heavy involvement in mining was at the Eldorado Mine at Uranium City. Working at the Eldorado took all the skills and talent I had — and then some. In fact, it was at the Eldorado Mine that the idea of Wirtanen Electric actually took hold. We founded our company in Edmonton in 1953 to meet the electrical and mechanical needs of large industrial projects starting up in Alberta. Eventually, we became a major electrical contractor in the province.

Together with my two sons, Wirtanen Electric has concentrated on maintenance, a motor-winding shop, some rentals and instrumentation. In 1972, we changed our business focus and concentrated on instrumentation work and rentals.

I first met Chuck Doerr in the early '60s, and helped him with the electronics of a giant bulldozer he was building. He also used my help and my instruments on a project auditing draglines. Although he was about 10 years older than me, I found Chuck a hard man to keep up to. He knew draglines and he could stand and look at one and tell whether it was up to capacity.

Not long ago, a very prominent electrical man came here and described our place as "the Aladdin's Treasure Cave of the electrical business." With that in mind, we're going to open a small museum in our shop area, and open it to the public.

Thomas Back in my day, in south Wales, coal mining was about the only occupation available. My father was a coal miner, and his father was a miner. I just followed the family tradition.

I started in Wales in 1914, just before the outbreak of the First World War. I was 13. Many older boys were going into the service, so mine owners were grabbing wee kids to work on the top of the mines. I was a chain packer. Then I went into the lamp house.

I came to Canada, and to Hillcrest, in 1927. My first job was bucking coal in the chutes.

By studying and working hard at various jobs, I became assistant mine manager in the West Canadian Collieries in Blairmore. I was manager of that mine for the last 13 years of its operation — until it closed in 1961. We knew for many years it was going to close — the markets were declining because of the switch to diesel fuel. I had more than 600 men working for me, and the company paid me only $550 a month. I worked 34 years in the Crowsnest Pass.

For a short while, I was pit boss at Star Key Mine, just north of Edmonton. Then I got involved in strip mining at Taber, Sheerness and Wabamun. My last shift was in 1972, at Grande Cache.

Being a mine manager was a grind and stressful most of the time, particularly in the latter years when the industry was becoming marginal. We lost about $4 a tonne during the last year, because we couldn't sell the coal to the railway or to anyone. I enjoyed it, but it was rough.

Thomas Morgan, Retired Coal Mine Manager

Sil Silica Sand Quarry, Bruderheim

The wind-blown silica sands of Bruderheim are part of the extensive Redwater Dune Field. The sands lie on the surface of this network, covered only by a thin layer of soil and vegetation.

Sil Silica acquired these sand dunes in the late '60s. Mining the site started out as a three-man operation, but the business has grown to employ 20 persons as its markets have expanded. Today, the product is shipped throughout western Canada.

When uncovered, the sands are easily dug with front-end loaders. Sent through an efficient pit washing plant where all clays and fines are removed, the product is then dried and screened. At this stage, the product averages about 92 per cent pure silica. The plant then sorts the sand into consistent, usable sizes.

Sizing is dictated by the end customer. The fibreglass industry uses a very fine mesh sand, while other companies use a heavier product for grouting mixtures or filtration. Sand ground to the consistency of flour is used for thermal cementing.

Terry Paisley, Quarry Supervisor

Doug MacLeod, Quarry Plant Operator

Matt Grimm, Superintendent and Rick Shewchuk, Plant Supervisor

Thorhild Coal Mine, Thorhild

Traditionally, local Ukrainian farmers in the Thorhild region dug coal for their own use from the shallow coal seam running through the area. Deposits from the seam were close enough to the surface to be found by farmers digging water wells.

In 1937, two brothers opened the Thorhild area's first coal mine on neighbour Alex Duleba's farm. Mike and Charlie Libicz brought in three experienced coal miners with helpers to dig an underground slope. This sophisticated and well-designed underground mine functioned until 1944, when the operators began to surface mine the coal. In 1950, the site was sold to John Meleshko who operated it until it was purchased by the present owner in the early '60s.

After a recent fire burned down the tipple, it seems that North Point Mine has permanently closed its doors. Although sufficient coal remains on the property, the owner says he has had enough of the difficulties in owning and operating a small mine. Continued operation would require construction of a new tipple to properly size the coal. In addition, Thorhild coal has to be properly stored because of its tendency to break down if stockpiled in the open air.

Traditionally, coal from North Point was sold as far away as Manitoba and Saskatchewan. Esoteric uses for the coal included the local custom of giving it to pigs for routing and chewing, although some farmers suggest this practice "doesn't do the pig much good." Others argue that the practice controls worms in pig populations. The village also used its coal to thaw frozen ground in the local cemetery during winter months.

Now that the region's lone coal mine is closed, the locals express regret at its loss, and even talk about getting another small mine started.

Louis Douziech, Operator

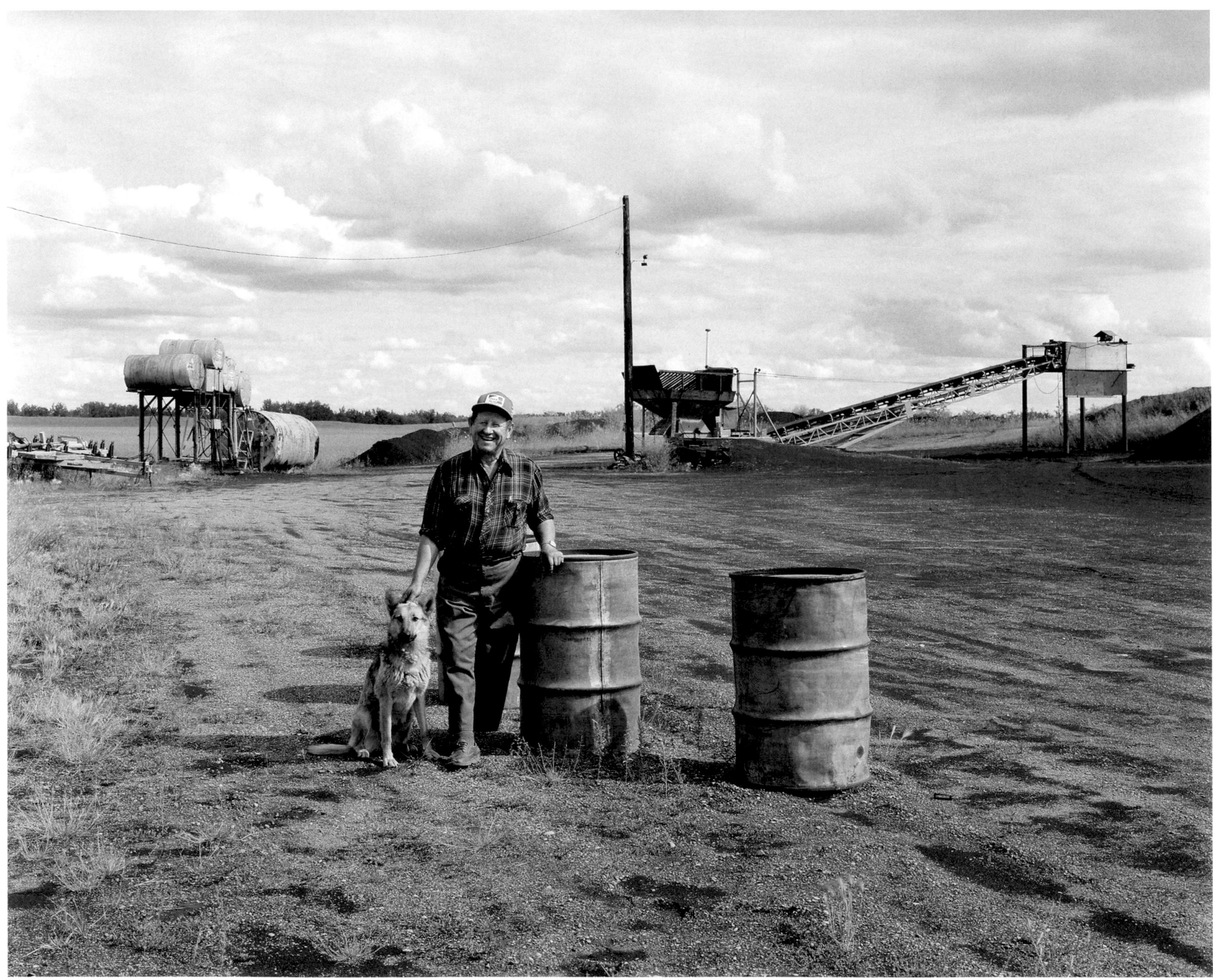

Dmytro Kolybaba, Surface Coal Miner

The names "Salt of the Earth," "Windsor Salt," and "The Canadian Salt Company Limited" are synonymous with Lindberg, Alberta. Evaporated salt was first produced at The Canadian Salt Works in 1948, on a small scale. Today, production capacity has increased to 400 tonnes per day. The Lindberg plant is the only salt producer west of Ontario and the principal supplier of the product in western Canada, although a similar operation once existed in Fort McMurray.

Rock salt is often mined using underground methods, but in Alberta, salt is extracted using the brine method. Holes drilled 3,700 feet below surface tap into a 300-foot-thick salt bed; fresh water is pumped underground and comes back as a concentrated brine.

In the plant, the brine is evaporated by steam heat in large, vacuum-operated evaporators. The product is then dried, conditioned and further processed into various products. For the Lindberg facility, most of the salt product goes to household, industrial, agricultural, water conditioning, and ice control use. The salt also acts as a basic raw ingredient for the chemical manufacture of soap, nylon, glass, oils and tars, as well as for meat curing.

The plant's smooth operation depends on continuous maintenance of its sophisticated equipment. Thus, 70 per cent of the employees are maintenance workers. The Lindberg Works has received several safety awards from the Salt Institute over the years.

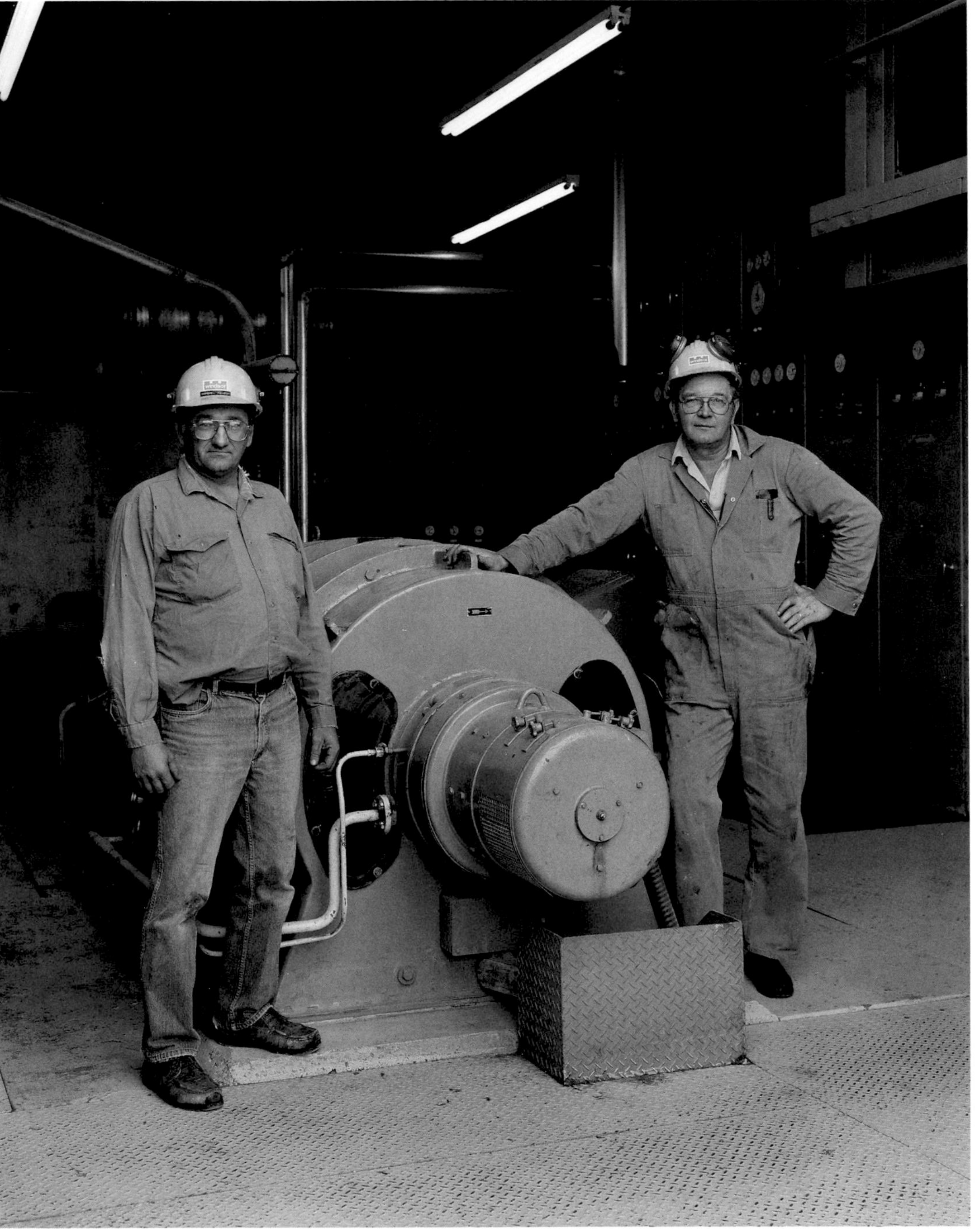

Marshal Pelech and William Pacholea, Steam Engineers

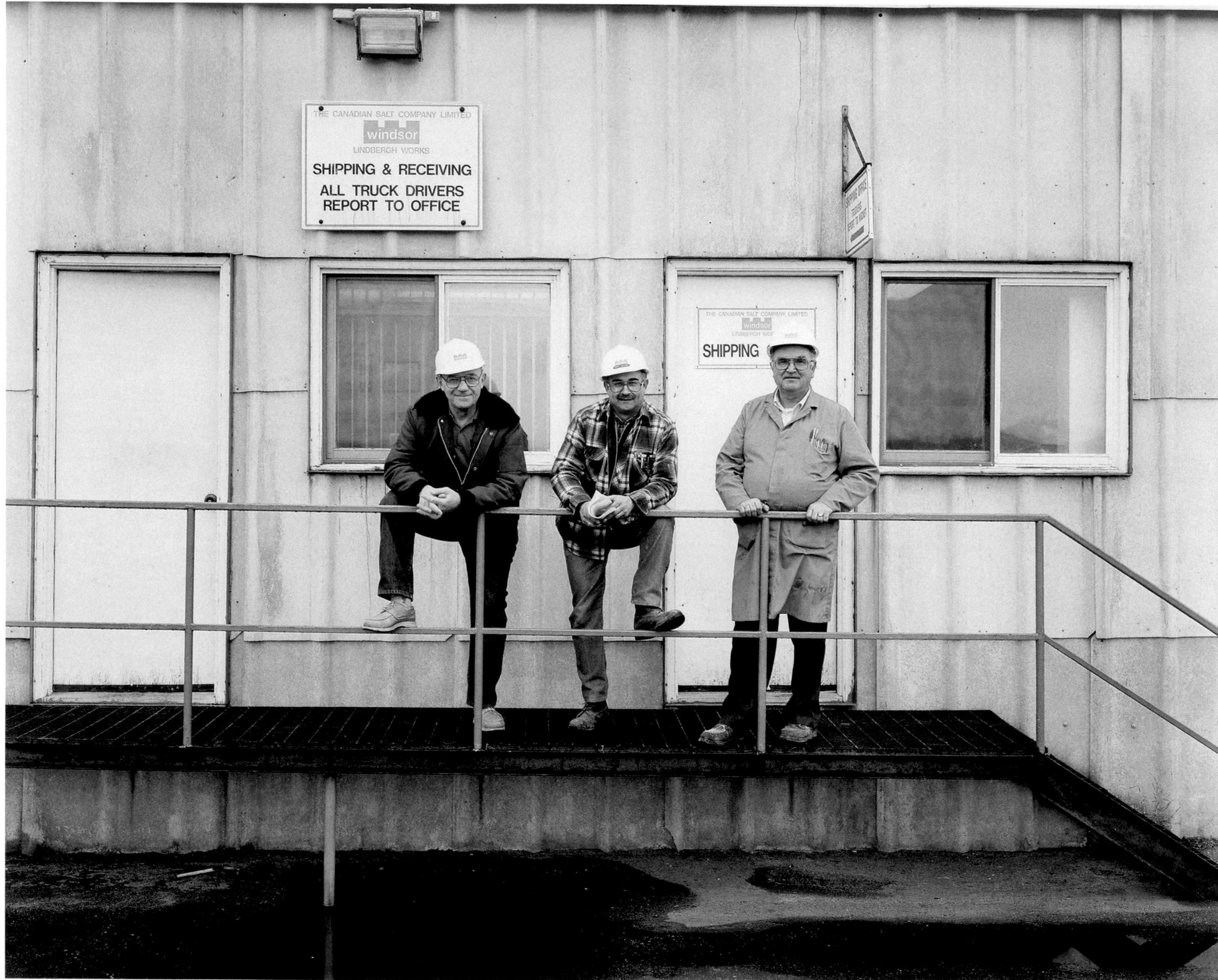

Mike Lotoski, Gary Gulayec and Walter Lesyk, Supervisors

Orest Fedorus, Joe Thir, Mike Breen and Lloyd Jenner, Maintenance Mechanics

Bill I was born right here in Fort McMurray, and so was my dad. He was a river boat captain, and never got involved in mining because he spent all his life on the river. When I started out, I spent six summers on the river and I couldn't see myself spending the rest of my life doing that. So I got out of it real quick-like. That didn't sit too well with my father because he wanted me to follow along behind him.

The rest of my life, I've been driving trucks and running equipment — that's all I've ever done. In 1964, I was living in Yellowknife and working underground at the Giant Gold Mine. But that's when they started building Suncor, and so I moved back to Fort McMurray. In fact, I was the first driver on the job — my number was #1. I worked those years for Bechtel — never officially for Suncor. The Suncor mine opened in 1967, and I've worked continuously at the Syncrude site since it started in about 1973. We were part of the crazy early days of the Fort McMurray construction boom.

My wife still works as a secretary for Syncrude out at the mine, but she's thinking of early retirement. As for myself, I usually get up in the morning and go down to the golf course. I'm getting right into golf. My handicap is down to 15, my best score is 80. I want to par that course one day — it'll probably never happen, but I like to think it could.

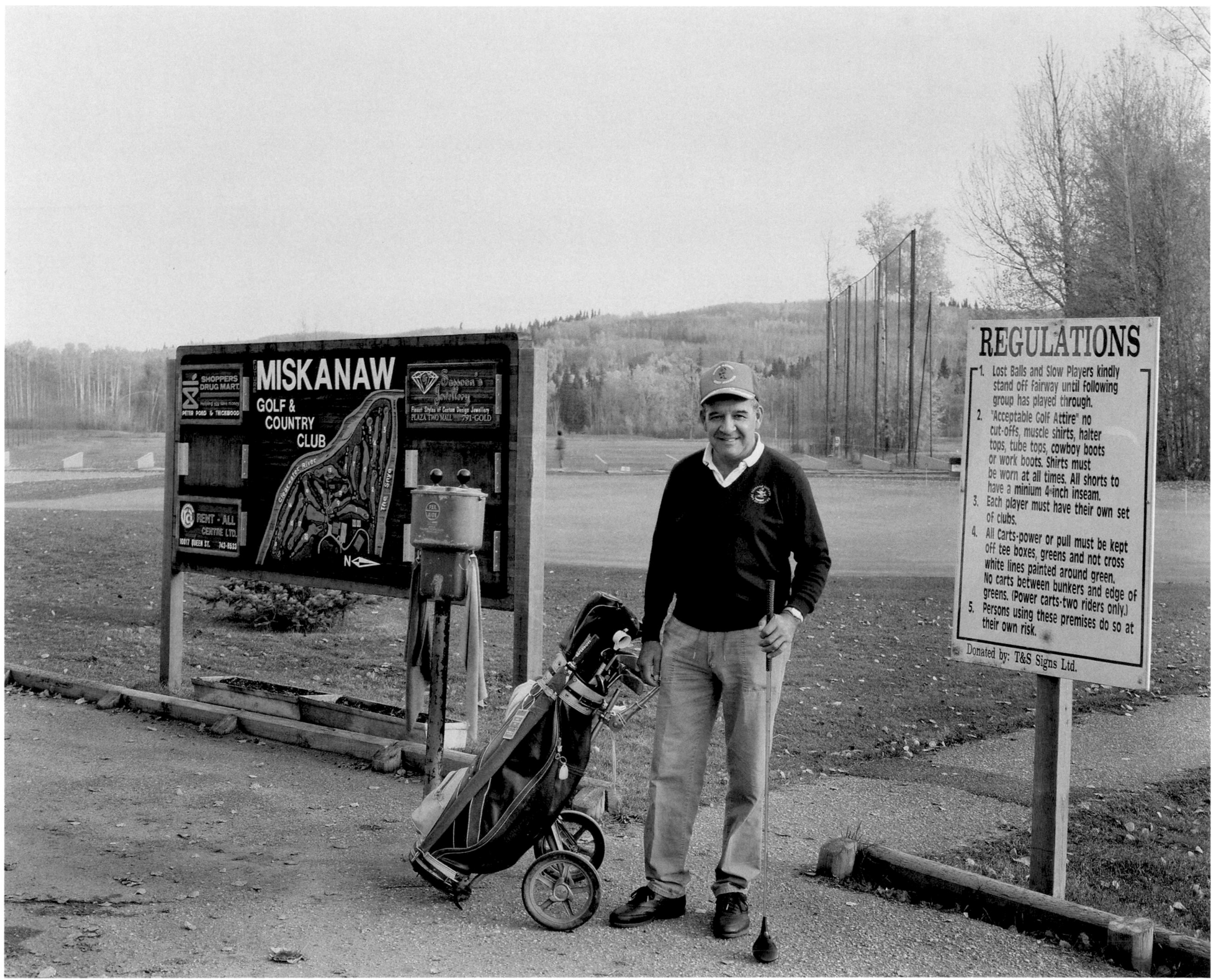

Bill Bird, Retired Oil Sands Worker

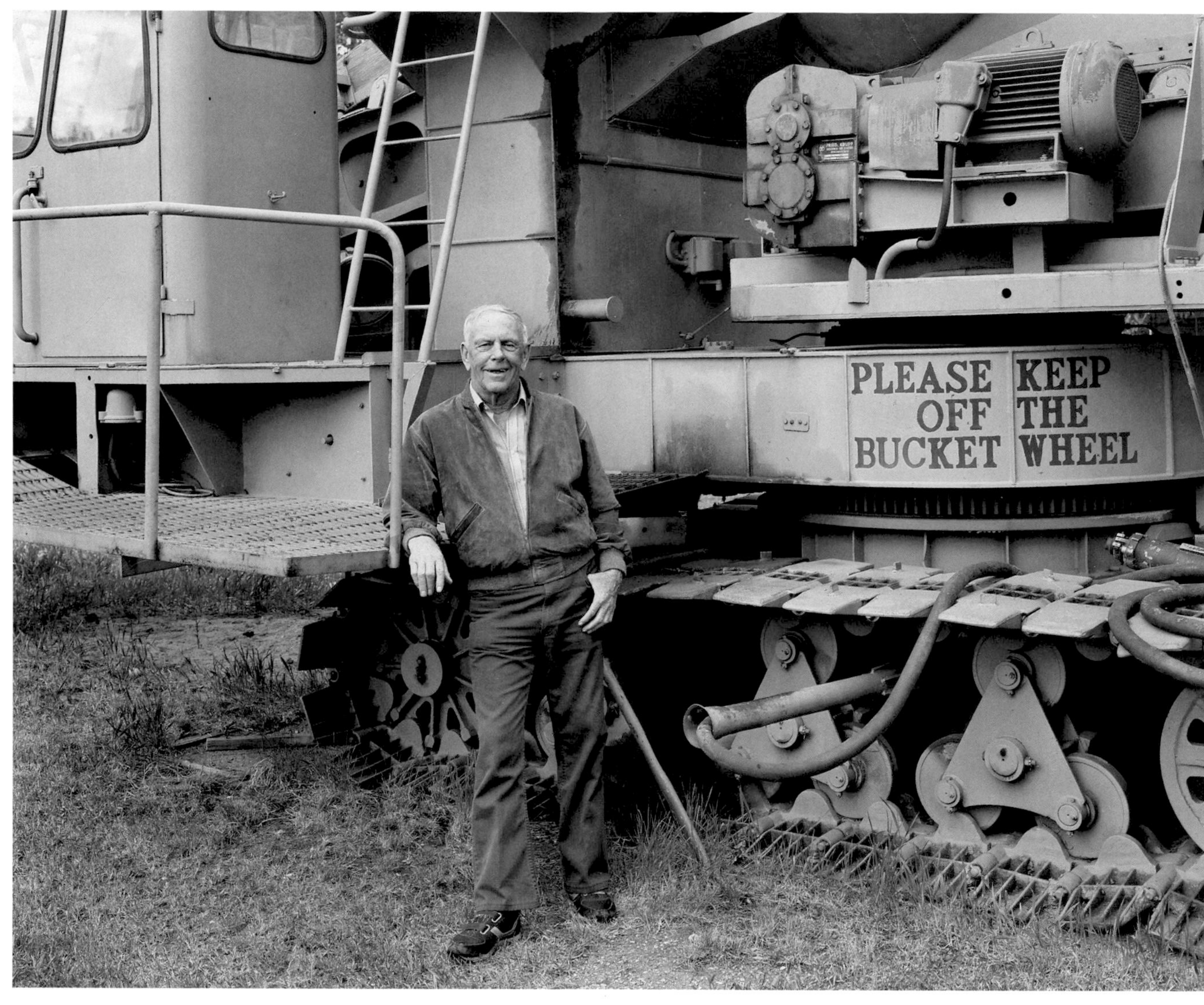

Glen McFadden, Retired Oil Sands Worker

Glen I was born in 1929 in Alliance, Alberta, just east of Camrose. My father was a farmer. In 1966, I was working for the Municipal District of Flagstaff on a grader, and I got a pamphlet in the mail advertising this place. It said to fill the pamphlet out if I was interested in working here, so I did, and two months later, I got a phone call.

I started as a grader operator with Great Canadian Oil Sands, which later was renamed Suncor. I worked on a variety of heavy equipment for Suncor.

Fort McMurray had a salt mine called Dominion Tar and Chemical. The salt was recovered as a brine from wells using water. They made the salt here in five and 50-pound boxes and salt blocks, and they shipped it all over Alberta and Saskatchewan

Now that I'm retired, I like to go fishing. I have a boat I take to Lac La Biche and the little lakes all around here. I think the fishing is getting poorer in this country. And I don't know who to blame for it — maybe it's because there are more fishermen.

Harold I was born in 1918 near Saskatoon. In 1938, we came to Fort McMurray to homestead at Wandering River, just 200 kilometres south. Twenty minutes after I got here, I landed a job in the bush for $28 a month, and I felt like a millionaire.

In 1939, I started with Abasand Plant, which had a little quarry and a plant. Abasand was intended to be an experiment, but they did make some oil and build a very small pipeline from the plant to Waterways — although I don't think they ever shipped anything.

In the early '40s, I worked for a few years at Bitumount, 90 kilometres north of Fort McMurray on the Athabasca River. Bitumount's oil supplies didn't amount to much either — they shipped it out a few barrels at a time on river barges. The operation had a small oil sands quarry nearby, but the main facility was for separation and refining the oil. Eventually, the government took over the plant. However, I understand that Bitumount helped develop the process for oil sands recovery that is used today by Syncrude and Suncor. It's an historic site now.

From 1960 to 1970, I worked for Cities Services until my job came under Syncrude control. Then I just moved across. In 1970, I went to Suncor and stayed there until I retired in 1983.

Most of my years were out in the field maintaining quarrying equipment. At Suncor, I ended as a millwright foreman, which involved doing major repairs on the bucketwheel excavators and conveyors.

I worked an awful long time with tar sands — it adds up to at least 27 years. And if I hadn't got to be 65, I would probably still be working. It was very interesting work, with some of the biggest and best equipment in the world — and good people. You can't beat them.

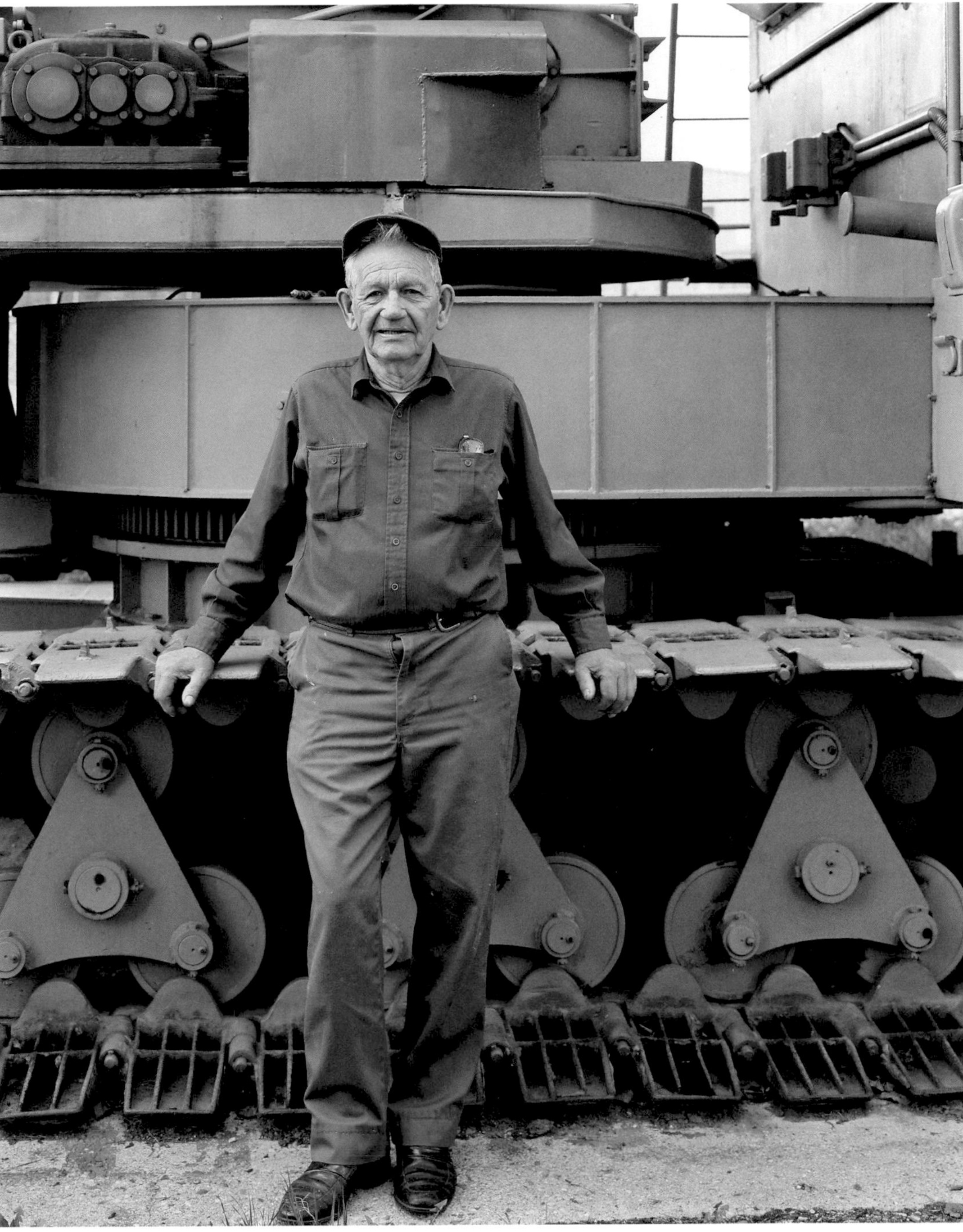

Harold Olsen, Retired Oil Sands Worker

James Coal mining never fazed me one little bit. It was a challenge to me, and if I do say it myself, I was good at it. It's the pride you take in it.

I was born in 1927 in Durham, England, the fourth generation of coal miners in my family. Originally, I had never wanted to go into the mines, and my father kept me out because he didn't want me to go underground with him and my two brothers, all of whom were working in the same colliery. Then, in the war years, the Essential Work Committee sent me to a coal mine. When I was 16, I was already underground. I ended up in the medical corps of the army, and got posted to Korea and Japan.

When the war was over, I returned to the mines and was there until 1966 when we decided to move to Elk Point, Alberta, where my sister was living. My wife used to hate it when I went in the pits.

At Elk Point, I worked at the salt plant in Lindberg for three years. We recovered the salt by pumping it up as a brine. The work was enjoyable, but the money was poor. In 1969, a fellow came around recruiting for Suncor in Fort McMurray. At first I started on the conveyors in the mine, but ended in the training department. I worked for Suncor until 1986, when I received early retirement.

Margaret and James Green, Retired Oil Sands Worker

When people think of oil, they normally think of oil wells.

But oil sands producers are not conventional explorationists requiring discovery dollars. The bitumen-rich sands of Athabasca are a known quantity — and their producers are miners, not diviners.

As their name implies, the Athabasca oil sands are huge ore bodies of compacted sand. The bitumen is generally found within 30 metres of the surface beneath bedrock muskeg and a layer of clay, silt and gravel.

In 1719, the Cree Indian Wa-pa-su guided the first explorers to the Athabasca oil sands, near Fort McMurray. It was not until 1778 that Peter Pond actually reported the presence of oil sand outcroppings along the Athabasca River.

In 1920, the Alberta government established the Research Council of Alberta to find an economic use for the oil sands. Ten years later, a pilot plant at Fort McMurray produced a small quantity of oil using the extraction methods of the Research Council's Dr. Karl Clark.

In the late 1960s, commercial quantities of oil were recovered from the Athabasca oil sands. Great Canadian Oil Sands, later to become Suncor Canada Inc., began commercial production in 1967 after four years of site and plant construction. A synthetic oil pipeline was built and houses were erected in Fort McMurray to accommodate the enormous influx of new employees.

In fact, when Suncor began construction, Fort McMurray was a mere village with a population of less than 1,000. Within 15 years, the village grew into a modern town of 25,000 people.

In the 25 years since Suncor's start-up, a total of four giant bucketwheel excavators have been used to mine oil sands. In 1991, Suncor produced a record 22 million barrels of synthetic crude oil from these known reserves. In future years, Suncor will introduce new and more competitive mining methods with a system of trucks and shovels to replace the gigantic bucketwheels of the past few decades.

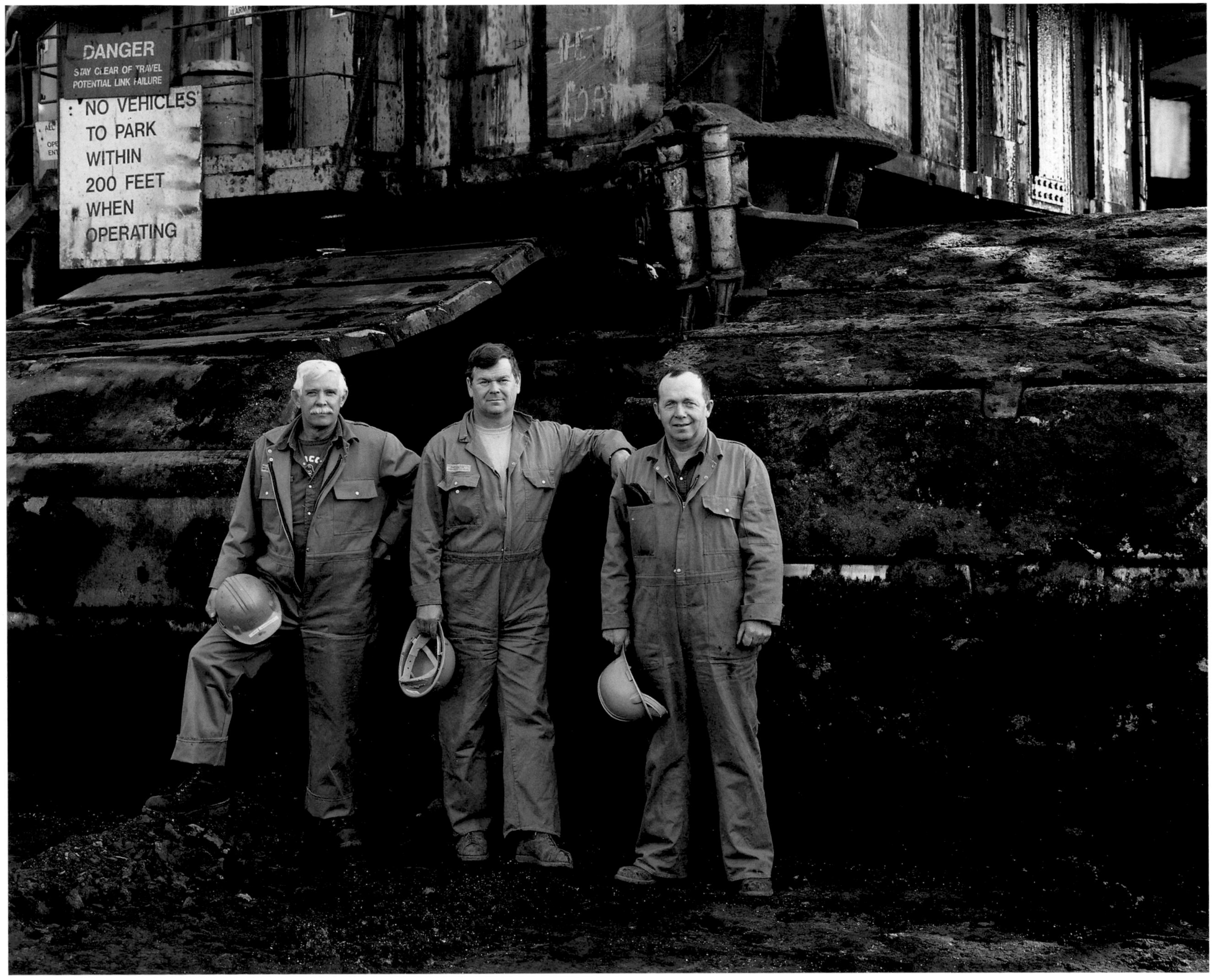
DANGER
STAY CLEAR OF TRAVEL
POTENTIAL LINK FAILURE
NO VEHICLES TO PARK WITHIN 200 FEET WHEN OPERATING

Art Morris, Tom Keating and Ken Partington, Oil Sands Workers

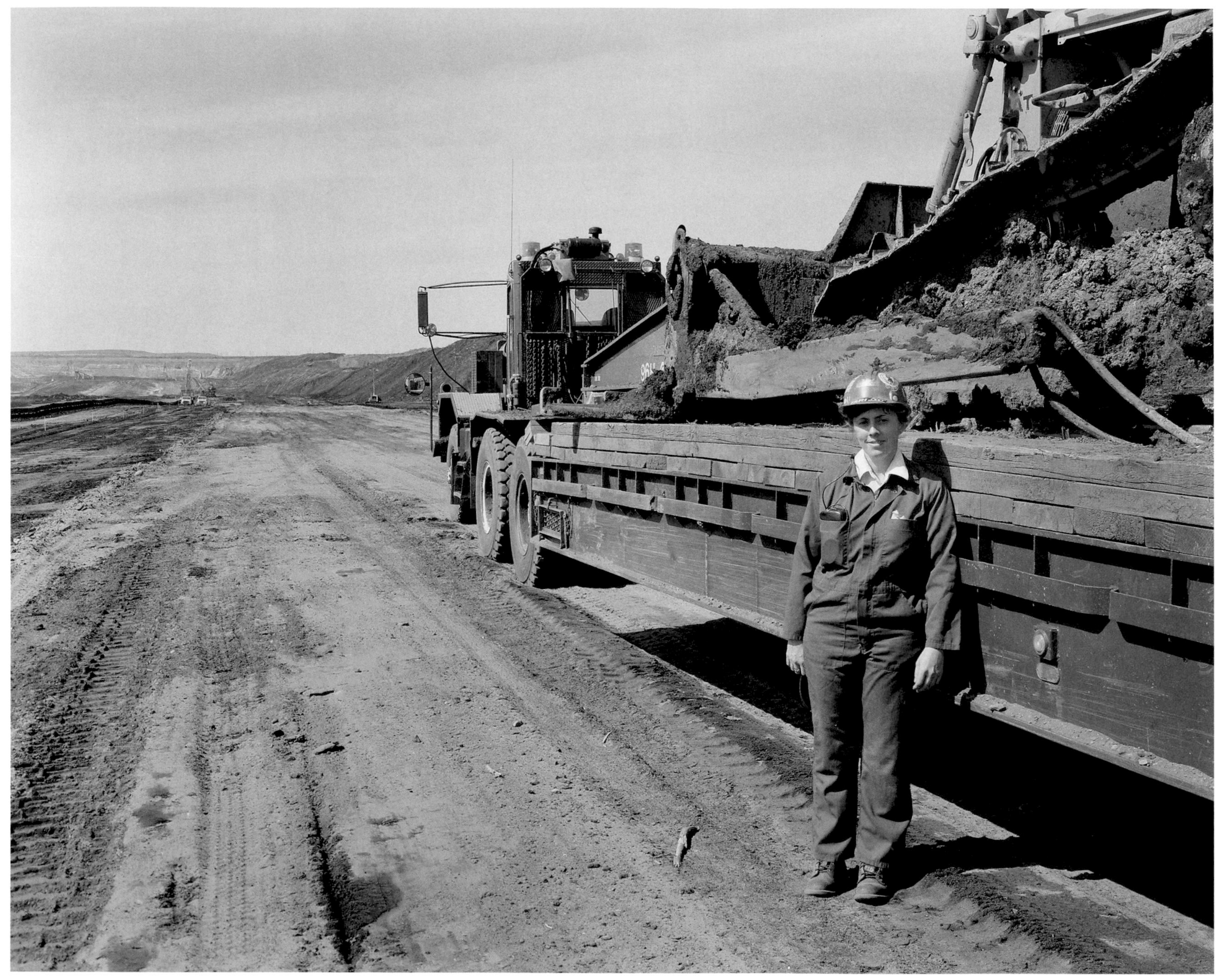

Alana Colwell, Heavy Equipment Operator

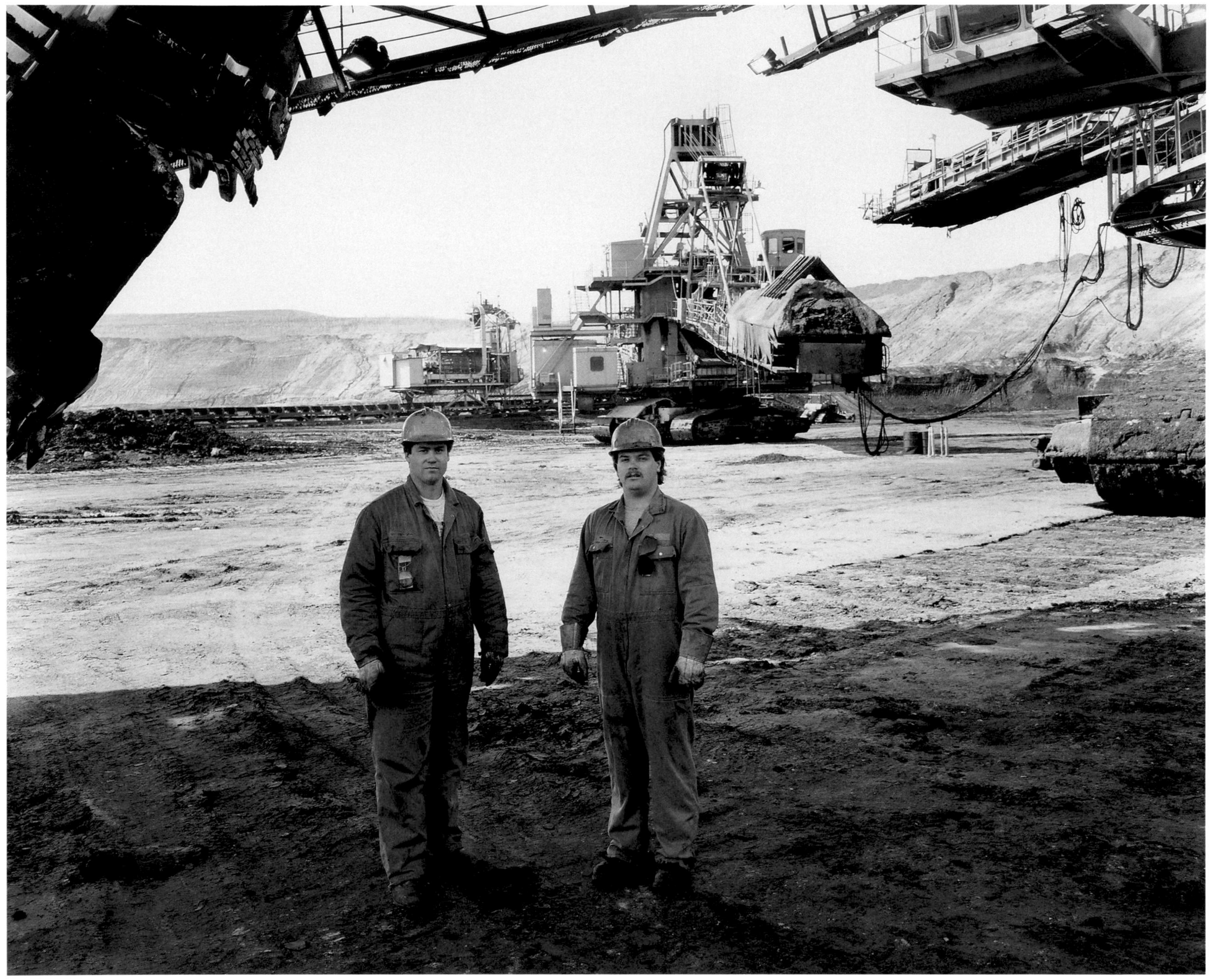

Blaine Kaiser and Rick Legge, Oilers

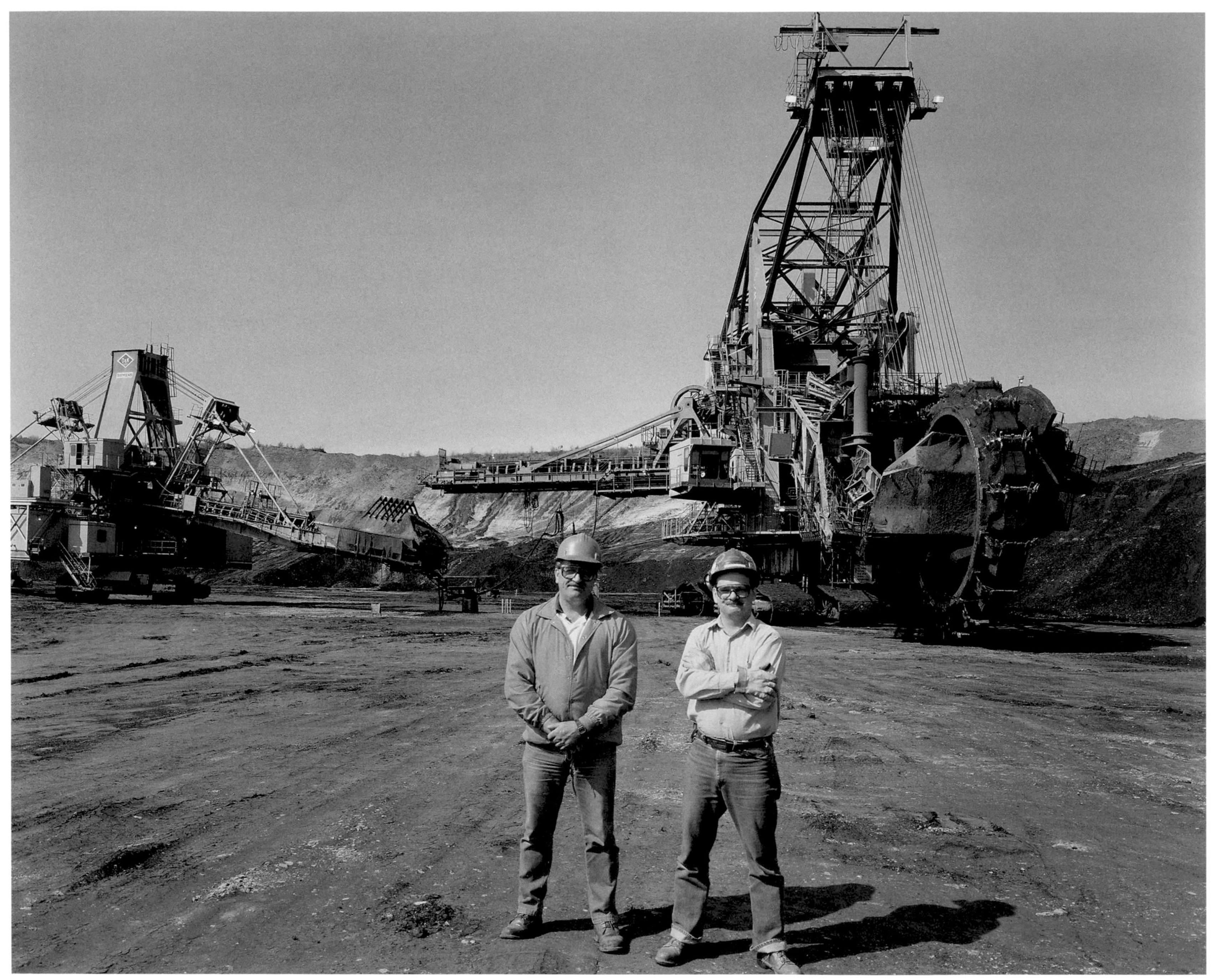

Don Monagher, Heavyduty Equipment Operator and David Ryan, Training Specialist

Jim Doodey, Electrician

Syncrude Oil Sands Mine, Fort McMurray

In 1973, after years of research and promotion, construction began on a second oil sands plant in northeastern Alberta. Designed to produce 80,000 barrels of crude oil per day, the Syncrude mine came on-stream in 1978 after a workforce of over 10,000 had completed this mammoth construction task.

By 1993, Syncrude was operating at more than twice its original design capacity, making it the largest producer of synthetic crude oil in the world (183,000 barrels of crude oil daily.) Today, approximately 1,400 people work at the mine, which is located 40 kilometres north of Fort McMurray. Another 2,900 people are employed elsewhere in the operation.

From its inception, Syncrude has maintained a team approach, or participatory management system in which people achieve company goals in an atmosphere of mutual respect and co-operation. Each mine production employee belongs to two teams: a work team and a shift team. For example, the mine's three bucketwheel operators are trained to operate the excavator as a team, sharing both responsibilities and problem-solving.

Syncrude workers are non-unionized, and are paid for the qualifications and skills they bring to their job as opposed to being paid a set rate for prescribed duties. Comprehensive training programs allow employees to continuously upgrade their skills and knowledge.

In addition, Syncrude has a unique two-year, on-site apprenticeship program for millwrights, welders, mechanics and steamfitters. Long-service employees develop personal career goals by working in various capacities throughout the mine.

Oil sands mining has some unique characteristics, many of which can be downright inconvenient for industry workers. When the weather warms, oil sands melt and tend to stick to everything, including workers' boots and equipment. When it rains, mine roads often become so impassable that in some cases, miners have been dropped by helicopter into their equipment at the pits.

Gord Anderson, Delores MacLean and Norm Young, Bucketwheel Operators

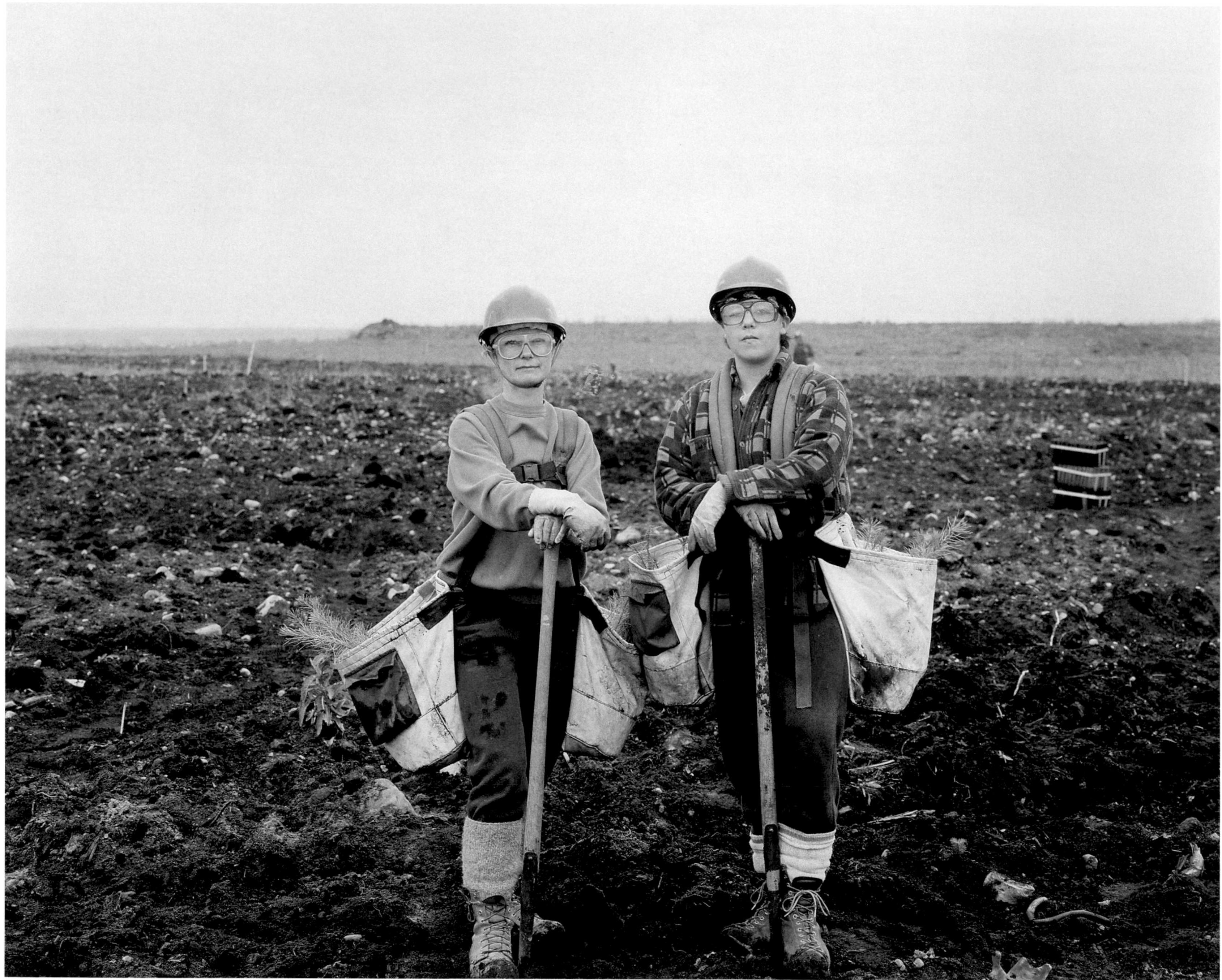

Gloria Couture and Ilona Garay, Tree Planters

Delores MacLean, Bucketwheel Operator

Syncrude Oil Sands Mine, Fort McMurray

Don Wheeler, Electrical Supervisor, Michael O'Quinn, Machinist and Ron Severson, Scheduler

Jeff Reid, Joe Joyce, Oscar Monplaisir, Michel Laberge and Mike Waddington, Millwrights and Machinists

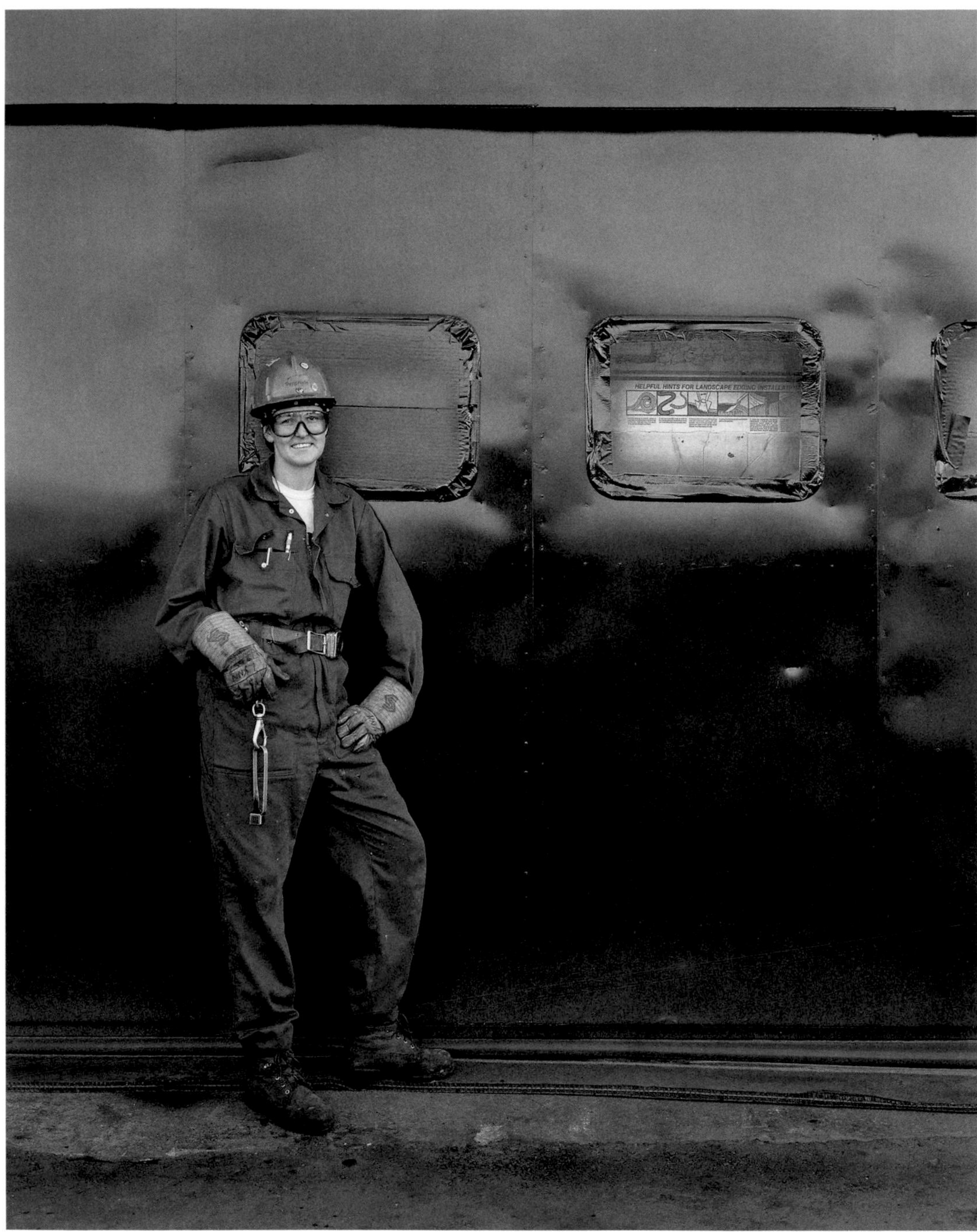

Lynn Smith, Welder

Joe Hamelin, Dragline Operator, Bill McPherson, Welder and Ivan Kosh, Dragline Oiler

Underground Oil Sands Test Facility, Fort McMurray

Although the Athabasca oil sands near Fort McMurray are one of the world's largest sources of bitumen, most of the deposits are uneconomical and too deep to mine using conventional surface methods.

In 1976, the Alberta Oil Sands Technology and Research Authority began investigations of *in situ* technology for the oil sands. The research resulted in a new underground mining concept: steam would be injected into the oil sand's layer through wells drilled from a tunnel just below the sands deposits. Steam-heated bitumen would soften and flow by gravity to an underground pipeline system, which would then transport the product to the surface.

In 1984, construction began on the first underground test facility for this experimental idea. Two 210-metre shafts were sunk in 1986, while in the limestone bed beneath the oil sands, large (five metres wide and four metres high) horizontal tunnels were dug outward from the main shaft to ease drilling holes for steaming and recovery.

Results from this initial testing led to a full-scale pilot test with drill holes reaching 600 metres in depth.

Miners on this Fort McMurray maintenance crew universally come from coal mining backgrounds, and all say that with its lack of dust and tedium, the oil sands facility is a "Cadillac of mines" — although one seasoned worker noted that a young person might find this job unfulfilling because "you don't see the bitumen that is finally produced."

Studies prove that conventional surface recovery methods can retrieve only six per cent of an oil sands deposit like the one in Athabasca. Methods perfected at this test facility could well represent the future of mining technology for the oil sands.

Derek Bell, Chris McCance and Bruce Holesworth, Maintenance Crew

David I started working in the mining industry in 1964, with England's National Coal Board. Later, I obtained a degree in mechanical design engineering and held a number of engineering and management positions there.

In 1981, I moved to Grande Cache where I became chief mechanical engineer with McIntyre Mines, and for the next nine years, held various positions in the coal mines in Cape Breton. In 1990, I joined Norwest Mine Services and the AOSTRA UTF project as mine superintendent.

Maurice I started out in England as a coal mine brick layer, then got into coal mining. I came to Canada in 1977 because a friend said he could get me a job at Smoky River. I stayed two years, then moved to the Lethbridge area to help sink the Kipp Mine shaft.

I came here in 1986 and enjoy my work as senior mine examiner — but it's a change from coal mining.

Colin I was born and raised in Fort McMurray. After high school, I went to Keyano College in town where I studied computer science.

During the summers, I got a job at the Underground Test Facility and in 1991, started here full time as an underground geo-technician.

John In 1954, I started coal-mining in southern Wales — I was 15 years old and at least the third generation coal miner in my family.

They advertised in the Daily Mirror that Grande Cache needed 20 coal miners, so I applied and got a job. That was in 1979. I worked at a few properties in western Canada, including Obed Mountain and the Luscar mines.

I came here in 1986 and helped break out of the shaft and drive the tunnels. Recently, I came back to work as a mine examiner — and it's got to be the best job I've ever had.

David Wright, Maurice Entwistle, Colin O'Brien and John Hooper, Mining Crew

Selected References

Appleby, Edna. *The Story of an Era,* Canmore, AB: Self-published, 1975.

Babaiian, Sharon, and Felske, Lorry. *The Coal Mining Industry in the Crow's Nest Pass,* Edmonton, AB: Historic Sites Service, Alberta Culture, 1985.

Blower, James. *Gold Rush,* Toronto, ON: Ryerson Press, 1971.

Caragata, Warren. *Alberta Labour — A Heritage Untold,* Toronto, ON: James Lorimer and Company, 1979.

Carroll, W.F. (Chairman). *Report of the Royal Commission on Coal,* Ottawa, ON: King's Printer, 1947.

Crabb, J.J. *Crowsnest Pass Travelog,* Calgary, AB: Crowsnest Resources, 1982.

Crowsnest Pass Historical Society. *Crowsnest and its People,* Coleman, AB, 1982.

Crowsnest Pass Historical Driving Tours. *Coleman, Blairmore,* and *Bellevue and Hillcrest,* Edmonton, AB: Alberta Culture, 1990.

Dowling, D.B. *The Edmonton Coal Field,* Ottawa, ON: Canada Department of Mines Memoir No. 8-E, Government Printing Bureau, 1910.

Drumheller Valley History Association. *The Hills of Home,* Drumheller, AB, 1973.

Energy Resources Conservation Board. *Coal Mine Atlas — Operating and Abandoned Coal Mines in Alberta,* Calgary, AB, 1985.

Fording Coal Limited. *Black Gold — A History in the Making,* Calgary, AB, 1991.

Fryer, Harold. *Ghost Towns of Alberta,* Langley, B.C.: Stagecoach Publishing, 1976.

Gadd, Ben. *Bankhead — The Twenty Year Town,* Calgary, AB: The Coal Association of Canada, 1989.

Hart, Hazel. *History of Hinton,* Hinton, AB: Self-published, 1980.

Hlady, Ernest. *The Valley of the Dinosaurs, Its Families and Coal Mines,* East Coulee, AB: East Coulee Community Association, 1988.

Johnston, Alex, and Gladwyn, Keith and Ellis. *Lethbridge: Its Coal Industry,* Lethbridge, AB: The Lethbridge Historical Society, 1989.

Johnstone, Bill. *Coal Dust in my Blood — The Autobiography of a Coal Miner,* Victoria, B.C.: Heritage Record No.9, British Columbia Provincial Museum, 1980.

Kerr, J. William. *Frank Slide,* Priddis, Alberta: Barker Publishing Ltd., 1990.

Manson, Jack M. *Bricks in Alberta,* Edmonton, AB: Alberta Masonry Institute, 1983.

Maydonik, Allen. *The Luscar Story,* Edmonton, AB: Luscar Limited, 1985.

Nordegg, Martin. *The Possibilities of Canada are Truly Great — Memoirs 1906 - 1924,* Toronto, ON: Macmillan of Canada, 1971.

Parks, Wm. A. *Report on the Building and Ornamental Stones of Canada,* Volume IV, Provinces of Manitoba, Saskatchewan and Alberta: Department of Mines, ON, 1916.

Patching, T.H. (ed). *Coal in Canada,* Special Volume 31, Montreal, P.Q.: The Canadian Institute of Mining and Metallurgy, 1985.

Ross, Toni. *Oh! The Coal Branch,* Edmonton, AB: Self-published, 1974.

Stephenson, H.G. and Luhning, R.W. *Oil at the End of the Tunnel,* Denver, CO: AIME, 1991.

Syncrude Canada Ltd. *The Syncrude Story — In Our Own Words,* Fort McMurray, AB, 1990.

Page Number

Southern Alberta

Central Alberta

Northern Alberta

Acknowledgements

The photographer acknowledges the generous support of The Alberta Chamber of Resources and the Canada Council toward the production of this book.

My involvement in this project has been greatly encouraged and bolstered by family and friends.

Special thanks is given to Donald Currie for his enthusiastic promotion of the book and his growing interest in large format black and white photography.

Alberta Culture and Manalta Coal provided assistance for an early part of the photography.

In addition, the mining companies have warmly facilitated my visiting their operations to photograph a cross-section of their employees.

Lastly, my thanks to all the men and women I photographed; I enjoyed sharing your stories. My only regret is that there was not space to include everyone.

Colophon

Design:	Darrell Robinson
Editing:	Kerry McArthur, Westword Communications Joan Rickard, Author Author Literary Agency Debbie Thomas, Lexicon Communication
Printing Consultant:	Norm Belanger, NJB Communications
Printing:	Kaleidocolor
Duotone separations:	Colour Four Graphic Services Ltd.
Binding:	North-West Book Company Ltd.

Biographical Notes

Lawrence Chrismas

When Lawrence Chrismas began photographing Canmore, Alberta coal miners in 1979, he had no idea he'd still be at it 14 years later. It was the beginning of a major photographic documentary of Canadian miners — coal miners, in particular.

Chrismas has compiled a veritable photographic history of coal mining in Canada (some of the older miners' recollections hail back to the 1920s), and probably the most extensive collection of coal miner photographs on the continent. To assemble these portraits he has travelled extensively to mining communities throughout Canada, particularly Alberta.

Born in Alberta and educated at the Universities of Alaska and Alberta in geology and mining, Chrismas has worked 25 years in the mining industry. He has held the position of Chairman of the Coal Division of the Canadian Institute of Mining and Metallurgy, and has served four years as the Chairman of the History and Heritage Committee of The Coal Association of Canada.

Chrismas has exhibited widely in Canada since 1978, and his photographs may be found in various public, private and corporate collections. His books include the 1989 *Manalta Miners* and *Minto Miners,* and he has also published numerous portfolios, catalogues and posters. His permanent installations are at the Minto Coal Museum, New Brunswick, and the East Coulee School Museum, Alberta; several others reside in corporate collections.

Thomas H. Patching

Thomas H. Patching, a resident of Edmonton, was Professor of Mining at the University of Alberta from 1947 until his retirement in 1980. He was Chairman of the Coal Division of the Canadian Institute of Mining and Metallurgy in 1967/68, and President of the Institute for the 1971/72 term.